SAFETY ANALYSIS
Principles and Practice
in Occupational Safety

SAFETY ANALYSIS
Principles and Practice
in Occupational Safety

Lars Harms-Ringdahl

Institute for Risk Analysis and Safety Management
Stockholm, Sweden

ELSEVIER APPLIED SCIENCE
London and New York

ELSEVIER SCIENCE PUBLISHERS LTD
Crown House, Linton Road, Barking, Essex IG11 8JU, England

WITH 47 TABLES AND 40 ILLUSTRATIONS
© 1987 & 1993 Lars Harms-Ringdahl

THIS ENGLISH EDITION
© 1993 ELSEVIER SCIENCE PUBLISHERS LTD

British Library Cataloguing in Publication Data

Harms-Ringdahl, Lars
Safety Analysis: Principles and Practice
in Occupational Safety
I. Title
363.11

ISBN 1-85166-956-6

Library of Congress CIP Data Applied for

Printed in Great Britain by Galliard (Printers) Ltd, Great Yarmouth

Preface

Over a number of years, great interest has been shown in the prevention of accidents that may have major consequences. This applies above all to technologically advanced installations in the chemicals processing and nuclear industries. A great deal of effort has been put in, and much research and practical work on development has been devoted to how major accidents can be prevented. Safety analysis has become a methodology that is applied to an ever greater extent, often providing the basis for safety activities at plant level.

Occupational accidents are another serious problem, even greater than major hazards. In the world as a whole, the International Labour Organization (ILO) estimates that around 180 000 people are killed and 110 million injured at work each year (Kliesch, 1988). In the light of the scale of the problem, this area deserves at least as much attention as that paid to major accidents. Safety analysis can also provide major assistance in preventing common accidents at work. The aim of this book is to describe some simple methods for analyzing hazards in the occupational environment. A safety analysis has three main elements: identification of hazards, assessment of the risks that arise, and the generation of measures that can increase the level of safety.

Safety analysis as a tool

The basic perspective of this book is that safety analysis is a tool that can be employed in safety work. By utilizing appropriate methods, the knowledge that is available at a workplace can be supplemented and made more systematic. A basic presupposition is that a systematic way of working is adopted.

Safety analysis is currently employed in the arena of occupational accident prevention to only a small extent. But, if the aim is to reduce occupational risks, methods which could be put to much greater use are available. Moreover, experiences from applying them are favourable. This applies both to identifying hazards and preventing the risks that arise from them. Furthermore, systematic safety work has economic benefits, for both companies and society. The increased complexity of modern production systems and the difficulties involved in understanding operations as a whole mean that a systematic approach to safety activities is called for. Otherwise, risks cannot be handled rationally and efficiently.

Safety analysis is one tool among others. It is not a shortcut but represents part of safety work as a whole. It is not particularly difficult, nor is it remarkable or peculiar in any sense. If an analysis is suitably designed, good results can be obtained from a few hours or a few days work.

Organization of the book

The aim of the book is to show how safety analysis can be practically applied in the field of occupational safety. Its emphasis lies on explaining how different methods work. The methods with a specific focus on occupational accidents are Energy analysis, Job safety analysis and Deviation analysis. To introduce the reader to a broader range of possible methods, a number of other approaches are also described. These include, for example, Hazard and operability studies and Fault tree analysis. The main focus is on qualitative methods, as it is the author's experience that quantitative methods have less applicability in this context.

Another major theme concerns the analytical procedure, i.e. the various stages that make up an analysis and how these are related to one another. This procedure includes the planning of an analysis. Some views are also expressed on how high quality analyses can be achieved.

Safety work and safety management represents a further theme. The aim of an analysis is to obtain results — in the form of a reduced risk for occupational accidents. An analysis should be seen as a supplement to a company's own safety activities. Arguments for and against conducting a safety analysis are presented in this context. One section is devoted to economic appraisals of safety analyses. Five examples are provided — as demonstrations that safety analysis tends to be profitable as well as socially desirable.

The book has been written primarily with safety engineers and the Labour Inspectorate in mind, but it should also be of interest to anyone involved in occupational injury prevention. The idea is that it will function as a guide for those who would like to employ the methodology. For this reason, there is no stress on theory; nor is an extensive bibliography provided. This is a revised and translated version of a book already available in Swedish (Harms-Ringdahl, 1987b).

The material is derived from both the specialized literature and the author's own work with the analysis and prevention of accidents. Certain important lessons have been learnt from training courses in safety analysis. The experiences of the course participants have led to improved accounts of methods and the various ways in which use is made of results. Experiences from these courses suggest that safety analysis is a procedure that should be more widely adopted, e.g. by safety engineers.

The word "company" comes up throughout the book, and this should be interpreted in the widest possible sense. It refers to any type of organized production where there is any type of physical risk. It does not imply a limitation to the private sector. The book is primarily concerned with occupational accidents, but the methods and approach can also be applied in many other areas and to different types of systems. There is some discussion of wider forms of application, referred to here as integrated approaches.

Those most interested in practical applications can skip over several of the chapters. In such cases, it might be appropriate to read the first six chapters and chapter 11, and quickly review the contents of chapter 12. A good way of getting underway is to start by looking at the account of systematic accident investigations in chapter 6. This will provide a deeper understanding of why accidents occur. Often it is only when you have tried yourself that the benefits of safety analysis become apparent.

Acknowledgements

The material in this book has been developed over several years through cooperation and contacts with many people. A start was made possible as a result of the interest shown by employees of the Swedish company, Fiskeby Board, where I was given the opportunity to test and implement a number of key ideas. An important role in the development of the book was played by the courses for safety engineers that took place at the Swedish Royal Institute of Technology during the early 80s. I would like to thank all those who taught on these courses, and am specially grateful for their capacity to clearly describe subjects which can only too easily become bogged down in detail. This particularly applies to Bo Bergman on reliability, Urban Kjellén on SMORT and planning of analyses, Carin Sundström-Frisk on psychology, and Juoko Suokas on HAZOP. My thanks also go to my research colleagues in Sweden and other countries, and to the many others who have encouraged the application and development of safety analysis in an occupational context.

I gratefully acknowledge the permissions to reproduce material from other sources. This applies in particular to the example of a company safety policy provided by the Association of Swedish Chemical Industries (sect. 13.1), the account of safety culture derived from the International Atomic Energy Agency (sect. 2.2), and the material on HAZOP supplied by the United Kingdom Chemical Industries Association Ltd. I also wish to thank Jon Kimber for his careful translation of the original Swedish text into English and Jonas Harms-Ringdahl for producing the illustrations.

Finally, I thank the Swedish Work Environment Fund for funding the translation into English and for the financial support they have provided for many of the projects on which this book is based.

Contents

1 Accidents and Safety

1.1 On terminology

Terms used in relation to accidents, such as hazard, risk, safety, etc., may have different meanings in different contexts. The meanings assigned to them largely depend on traditions that have been established within different academic disciplines and applied in a variety of technical contexts. It seems scarcely credible that a broad consensus on terminology can be reached in a field of study with such a wide spectrum of applications.

Some of the basic terms used in this book are discussed in this chapter. Neither a comprehensive review of the various definitions nor a major search of the literature has been attempted. Some references are provided, but these should be regarded as examples rather than as giving a definitive account of usage.

Accidents and incidents

An accident is an undesired event that causes damage or injury. An incident is an undesired event that might have caused damage or injury. The term near-accident is often used to describe the latter type of event.

One specific term is major accident, which is defined as "an occurrence such as a major emission, fire or explosion, resulting from uncontrolled developments in the course of an industrial activity, leading to a serious danger to man, immediate or delayed, inside or outside the establishment, and to the environment, and involving one or more dangerous substances" (CEC, 1982).

There are, however, other definitions which are frequently employed. In the medical tradition, the term injury is preferred to "accident" (Andersson, 1991). In such cases, "accident" means "an event that results or could result in an injury" (Karolinska institutet, 1989).

Hazard

The term hazard is often used to denote a possible source or cause of an accident. A somewhat different definition is that a hazard is an object or situation which constitutes a threat of loss (Lees, 1980). Source of risk has been proposed as an alternative term (SCRATCH, 1984).

Risk

The word risk is used in a variety of contexts and in a variety of senses. In general, it can be defined as the possibility of an undesired consequence, but is often regarded as a function of probability and consequence. In everyday speech, its meaning shifts between these two senses. A LARGE risk may refer to the seriousness of the consequences of an event occurring, or the high probability that it will occur, or a combination of the two.

In many contexts, risk is used to mean the likelihood of damage or injury. For example, risk is defined as the "probability of loss occurring" (Lees, 1980), or the "likelihood per unit time of death, injury, or illness to people" (Okrent, 1986). The term risk may also be used when outcomes are uncertain. Vlek and Stallen (1981) have listed definitions of objective risk which are common in the literature:

— Risk is the probability of a loss.

— Risk is the size of the possible loss.

— Risk is a function, generally the product of probability and size of loss.

— Risk is equal to the variance of the probability distribution of all possible consequences of a risky course of action.

— Risk is the semivariance of the distribution of all consequences, taken over negative consequences only, and with respect to some adopted reference value.

— Risk is a weighted linear combination of the variance of and the expected value of the distribution of all possible consequences."

This is a statistician's view on risk. Perceived risk may be something quite different. According to Brehmer (1987), "the most useful approach to psychological risk may well be to consider risk judgements as intuitive value judgements which express a diffuse negative evaluation of a decision alternative, a general feeling that this is something one does not want."

2

The multiplicity of ways in which the concept of risk is used does constitute a problem. In this book, it will be used in its general sense: the possibility of an undesired consequence.

Some examples of types of risks are provided below:

A. *Accident with direct consequence.* A sudden undesired event that is triggered off unintentionally and apparently by chance. The unfavourable consequence is observable within a short period of time. Examples include an accident where someone is crushed in a press, an explosion and the breakdown of an installation.

B. *Accident giving increased probability for injury or damage.* This is the same as A, but the consequence is not direct. An example is the increased likelihood for cancer arising from exposure to radiation or chemicals when an accident occurs.

C. *Slow deterioration or degeneration.* Examples include occupational diseases or environmental destruction caused by continuous exposure or the absorption of repeated small doses of chemical substances, prolonged overexertion, etc.

D. *Sabotage.* A negative event caused by the wilful action of a person.

The lines of demarcation between these categories can become blurred. For example, the difference between A and C is a question of time. In the case of A, it is a matter of fractions of a second to, perhaps, a number of hours. For C, it is often a matter of years.

This book focuses on sudden, undesired events. Its emphasis is on occupational accidents (those that occur at work), but its philosophy and manner of proceeding is applicable in principle to most risks of types A and B.

Safety

In a sense, safety is the opposite of risk. For this reason, an appropriate definition of a safe system is one that is free from obvious factors that might lead to injury of a person or damage to property or the surroundings (SCRATCH, 1984).

3

1.2 The problem of accidents

From a global perspective, accidents are a major health problem. Each year, there are nearly three million fatalities resulting from accidents or poisoning, of which two millions occur in less developed countries (Karolinska institutet, 1989). According to the same source, injury is the primary cause of death among children and young men in virtually all countries. The medical, social and lost productivity costs of all injuries are estimated to exceed 500 000 million US dollars each year.

In the USA, an annual total of 4.1 million life years are lost as a result of accidents and injuries (Committee on Trauma Research, 1985). The corresponding figures for heart disease and cancer are 2.1 and 1.7.

Accidents and injuries are the most common causes of death of Swedish men up to the age of 40 (NCBS, 1991a). Accident victims are often young people, meaning that they give rise to a relatively large number of lost years of life. In all, this means that "accidents are the most important cause of lost years of life below 65 years" (Andersson, 1991).

Risks on an individual level

People in society engage in a large number of activities which are hazardous to a greater or lesser extent. The level of risk varies considerably from activity to activity. Perspectives on risks at an individual level depend on a range of factors (e.g. Fischoff *et al.*, 1981). In general, higher risks are accepted in voluntary activities where the individual has a certain degree of control over what is happening. Also, they generally get something from what they are doing. Examples include mountaineering and motor cycling.

For many activities, however, risks are not taken on voluntarily, nor does the individual have much control over them. In such cases, higher safety requirements are demanded for risks to be regarded as acceptable. One example is that of air traffic, which is subject to extensive and detailed safety regulation. Another concerns occupational risks, which have to be kept as low as possible.

Table 1.1 lists the types of accident events that led to fatalities in Sweden in 1988 (NCBS, 1991a). It shows that men are more likely to be victims of accidents than women. The difference can be explained in terms of dissimilarities in types of work and other activities. It is also generally assumed that men tend to take greater risks.

Table 1.1 *Fatal accidents in Sweden in 1988 (population: 8 millions)*

Type of accident	Number	Men
Falls	1 254	45%
Motor vehicles	820	69%
Suffocation	150	57%
Poisoning	141	79%
Drowning	107	81%
Fires, explosions	99	72%
Water transport	71	96%
Aircraft	24	83%
Machinery	30	97%
Electrical current	13	85%
Lightning	1	0%
Other	309	—
Total	3 019	60%

Risks at company level

Companies face an extensive and diverse risk panorama. In addition to various commercial risks, a number of specific examples can be provided:

— Occupational injuries.

— Fires and explosions.

— Damage to machinery and equipment.

— Transportation injuries and damage.

— Product liability and damages.

— Harm to the environment resulting from the company's activities.

— Sabotage.

1.3 Occupational accidents

Occupational accidents are also a major problem from a world perspective. According to one estimate, 180 000 people a year die from accidents at work, while 110 million are injured (Kliesch, 1988).

In a large number of countries, both industrialized and less developed, the frequency of accidents resulting in a fatality has fallen since the 1960s. For example, it has fallen by 70% in Japan and Sweden, and by 62% in Finland (Kliesch, 1988). Similarly, the frequency of serious injuries is also falling, at least in the industrialized countries.

The explanations usually provided for this are that there are fewer people with hazardous occupations and that workplaces have become safer. However, a study of changes in accident frequencies for various occupational groups in Sweden between 1960 and 1980 presents a rather different picture (Fång, 1988). The people investigated worked in the wood products industry, slaughter-houses, metallurgy plants, mines and engineering workshops. In all cases, the accident frequency either remained constant or had even risen. These results demonstrate that the hypothesis that workplaces have become safer is not universally correct.

Some definitions

Occupational injuries can occur in a variety of ways. In general, they can be divided into three categories:

— Occupational accidents — accidents occurring at the workplace.

— Occupational diseases — harmful effects of work that are not due to an accident, such as overexertion injuries, allergies or hearing complaints.

— Commuting accidents — accidents occurring on the way to or from the workplace.

By an occupational accident is meant a sudden and unexpected event that leads to the injury of a human being in the course of his or her work. Generally, the course of events is rapid — lasting seconds or even less. But some, such as those involving poisonous gases or cold, might require several hours of exposure before an acute injury is incurred.

In order to make comparisons, some kind of measure of relative frequency is required. The measures vary between countries, often making it difficult for international comparisons to be made. One is the number of accidents relative to a given number of people employed, either 1 000 or 100 000. Alternatively, the frequency of accidents can be seen in comparison with the number of hours worked, usually 1 million. Statistical reporting also requires a specification of what is to be counted as an occupational accident. In Sweden, cases which lead to an absence from work of one day or more are included in the official statistics. Other countries employ different definitions, e.g. accidents leading to an absence of three days or more.

Other measures employed include number of days of sick leave per accident and days of sick leave resulting from accidents per employee. One difficulty with presenting the data in this way is that fatal injuries and those that cause disability have to be allocated a number of equivalent days.

Examples

A picture of the extent of the occurrence of accidents and other occupational injuries can be obtained from official Swedish statistics (NCBS, 1991b). It should be noted that the Swedish statistics include all injuries that lead to an absence from work of one day or more, which will affect comparisons with other countries.

In Sweden the number of fatal occupational accidents fell from 13 per 100 000 full-time employees in 1955 to 2.6 in 1989. The total number of reported injuries for 1989 was 176 000 for a working population of 3.8 million. Of these injuries, 59% resulted from accidents, 31% from diseases, and 9% from journeys to and from the workplace.

Table 1.2 *Frequency of occupational accidents in selected branches of industry in Sweden in 1989 (accidents per 1 000 employees)*

Branch of industry	Frequency	Man years/accident
Slaughter-houses	135	7
Iron works	122	8
Concrete products	90	11
Fire services	77	13
Saw mills	76	13
Docks (stevedoring)	75	13
Mining/mineral extraction	61	16
Construction	49	20
Manufacturing	39	26
Chemicals	36	28
Banking and insurance	3	312
Average	26	38

Table 1.2 provides examples of accident frequencies in various branches of industry for 1989. The distribution of risks is unevenly distributed. On average, slaughter-house employees run a risk of sustaining an injury that is five times as great as that of workers in general, and 45 times greater than bank or insurance employees.

Relative frequency can be a rather crude measure. All employees are included in the calculation, but it is almost only those who are directly involved in production who are at risk. A better picture is obtained by adjusting for the proportion of white-collar workers. Note that this adjustment has not been made in table 1.2.

How can the measure of frequency be explained in more practical terms? The table shows the number of man years per accident. The average frequency of 26, for example, is translated into one injury per employee every 38 years. This can be interpreted to mean that on average a person incurs an occupational injury once during his or her working life.

Table 1.3 shows the most common types of accidents. However, if fatal accidents alone were considered, the distribution would be considerably different. During the period 1980 to 1989, moving vehicles accounted for 40% of fatalities and "falling from a height" for 16%. The parts of the body most often affected are the hands (32%), followed by the feet and legs (23%), the back (16%) and the skin (13%).

Table 1.3 *Distribution of occupational accidents in Sweden in 1989 by type*

Type of accident	% share
Overexertion of a part of the body	22
Injured by handled object	15
Contact with moving machine part	11
Falling on same level	9
Struck by falling object	7
Contact with stationary object	6
Misstep, etc.	6
Falling to lower level	5
Struck by flying object, spray, etc.	5
Accident with moving vehicle	4
Other	10

There are data missing from the statistics. Some studies have concluded that a relatively large proportion of accident-related injuries are not reported as occupational injuries (e.g. Schelp & Svanström, 1986). The real number of accidents that lead to injury may be 30% higher, or even higher still.

Perceived risks at work

In one survey (NCSB, 1982), a question was posed to employees on how they perceived the accident risks they faced at work. 20% of all employed people considered that their work was either quite or very dangerous. 35% of skilled workmen regarded their work as hazardous. The highest proportion was recorded in the construction industry, where the figure was 51%.

From the perspective of the individual, perceived risks for accidents have great importance in terms of welfare. Naturally, awareness of the danger of being killed, disabled or injured has a negative effect on well-being.

In one sense, experiences of accidents are rare for the individual. Considerable periods of time elapse between occasions of injury. Nevertheless, on average an industrial worker will sustain an occupational injury twice or three times during his or her working life.

On the costs of accidents

Occupational accidents are also of economic importance — for society, for the company and for the injured person.

At a societal level, the costs are considerable, but difficult to discern and calculate. They affect different parts of the health care system, the insurance industry, etc. The European Community has estimated that the annual cost of occupational accidents and diseases amounts to 20 thousand million ECU. In Austria, it has been estimated that the economic cost of accidents at work comes to 25 thousand million schillings, equivalent to 2 thousand million US dollars (Thiel, 1988). For traffic accidents, the corresponding figure was 38; for accidents away from work, it was 40 thousand million schillings.

At company level, costs are also incurred, their significance depending

9

on which types of insurance and compensation systems are being operated, how sensitive production is to disturbances, etc. Heinrich (1931) and Broody *et al.* (1990) found these costs to be relatively high. On the other hand, in a report on the furniture industry, Söderqvist *et al.* (1990) stated that the marginal cost of accidents to companies was very low. Reasons for this were that comprehensive insurance policies covered compensation to injured persons, insurance premiums were independent of the number of accidents, and there was a certain surplus of personnel.

In general, it can be said that the total cost of accidents varies considerably between companies. The costs that are incurred can be divided up into the following categories:

— Insurance premiums.

— Compensation payments, care and rehabilitation of the injured person.

— Increased production costs, etc.

— Demands for safety measures and routines.

Examples of the costs related to production are as follows:

— Destruction of equipment and material.

— Interruptions to production, e.g. for accident investigations or repairs.

— Lower productivity and poorer quality in the short term. The replacement worker may be less skilled than the injured person, or work in general may be disrupted.

— Costs of recruiting a replacement and/or overtime payments.

— Indirect disturbances in that personnel may be emotionally affected by the accident.

— Costs incurred in investigating the accident.

— Specific demands for changes as a result of the accident.

At a personal level, an accident can create difficulties for the individual affected in a large number of self-evident ways. There is, however, a long tradition that people injured at work receive compensatory insurance payments.

2 Features of systems and accidents

2.1 Elementary reliability theory

The theory of technical reliability is really beyond the scope of this book. Nevertheless, as technical reliability has great relevance for system safety, a short account of reliability theory is provided here. Otherwise, the reader should refer to the more specialized literature (e.g. Lees, 1980; O'Connor, 1991).

Some concepts

Sometimes, it can be desirable to know the probability that a certain system will continue to operate for a specified period of time. Reliability of this sort is specified in terms of function probability or probability of survival and is usually denoted as $R(t)$, where R stands for Reliability and t is the length of the time period. The failure probability is the probability that the equipment will break down before the expiry of the time period t. This is usually designated as $F(t)$ and is related to $R(t)$ by the formula:

$$F(t) = 1 - R(t) \qquad (1)$$

The failure density shows the probability of the occurrence of the first failure (after time point 0). It is designated as $f(t)$ and is the time derivative of the failure probability. Other concepts are Mean Time Between Failures (MTBF) and Mean Time To Failure (MTTF). These two concepts are similar to each another, but not identical. MTBF only has meaning when it is applied to a population of components or systems where repairs take place (Lees, 1980). It is a mean value, derived by dividing total operating time by the number of failures. By contrast, MTTF is used for systems that are not repairable. It is also a mean value, but the mean of the distribution of the times to failure.

Depending on the types of failures that can arise and how the systems are constructed, different statistical models can be applied. Sometimes, the exponential distribution is used. On other occasions, it is the normal distribution or the Weibull distribution.

11

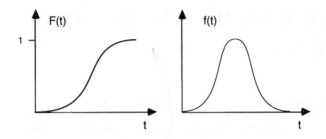

Figure 2.1 *Example of a failure distribution function and a failure density function (arbitrary time scale)*

A further important concept is that of the failure rate, often denoted as $z(t)$. It can be said to express disposition to fail as a function of time. Mathematically, it is defined as $z(t) = f(t)/R(t)$. Figure 2.2 shows a typical example of a failure rate function, known as the bathtub curve. The failure rate is high at the beginning, perhaps because of the presence of a number of weak components. These are replaced by components that function normally, resulting in a better-functioning system. The increase in the failure rate at the end of the time period is due to the wearing-out of the system.

Series systems

Technical systems usually comprise a number of subsystems, all of which must function for the system as a whole to work. These are to be linked in series. Let us suppose that the various subsystems have reliability

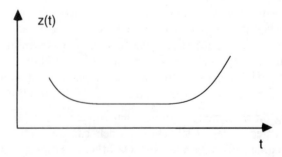

Figure 2.2 *The bathtub curve failure rate function*

probabilities of $R_1(t)$, $R_2(t)$, $R_3(t)$ etc. and that the reliability for the system as a whole is denoted as $R_K(t)$. We then obtain the relationship:

$$R_K(t) = R_1(t) \cdot R_2(t) \cdot R_3(t) \qquad (2)$$

If the probabilities of failure are small, an approximate expression for the failure function for the complete system can be obtained from a simple series development:

$$F_K(t) = 1 - R_K(t) = F_1(t) + F_2(t) + F_3(t) \qquad (3)$$

Parallel systems

In some cases, a system can be designed so that only some, but not all, of its components need to function for satisfactory operation to be achieved. One example is a light fitting with several light bulbs: for the fitting to provide no light at all, all the bulbs must fail. Another is a vehicle braking system with two independent hydraulic circuits: both must fail before there is no braking effect.

In terms of reliability, such systems are described as parallel or redundant, meaning that all the subsystems in question must fail before system failure occurs. The failure function for the system as a whole depends on the functions for the subsystems and is given by the equation:

$$F_K(t) = F_1(t) \cdot F_2(t) \cdot F_3(t) \qquad (4)$$

Calculations and data on failure

The use of reliability techniques gives rise to a need for data on component and system failures, and also information on times taken to make repairs. Data on certain types of human error may also be required. The data can be obtained from data banks or the technical literature, or may be collected directly. Where data are not available, estimates can be made. The level of accuracy needed largely depends on the application of the analysis.

The calculations may require advanced mathematical techniques. Moreover, many difficulties arise in obtaining and evaluating the data. One problem is the rapid rate of technological development. New versions of components emerge at such short intervals that the time required to evaluate their reliability is not sufficient.

Simple safety systems

In many cases, a component failure can be detected more-or-less immediately, e.g. when it has the direct effect that the system ceases to function. In other cases, the defect is not noticed until the particular component needs to be used.

Let us take a simple safety system as an example. Some hazardous machine parts are shielded by safety guards over inspection hatches. To prevent the machine from being run without the safety guards in place, each hatch is equipped with a safety interlock which stops the machine from operating when the guard is not in place. Certain types of failures may mean that the machine will not stop when the hatch is opened. This increases the probability of an accident.

One type of failure which is not dangerous by its very nature is when the machine stops despite the hatch being closed (fail-to-safe). But the stopping of the machine means that production cannot continue until the defect is remedied. This may give rise to a "second-order" risk, as there is a temptation to by-pass the safety system until the fault is repaired. This can take time.

A further example is a water tank heated by an immersion heater, which is controlled by a thermostat. One failure that is dangerous is when the thermostat permanently indicates that the temperature is low. This is a direct cause of over-heating and may cause an accident if the defect is not discovered.

Strategies for improving reliability

There are a variety of different strategies that can be applied and combined to improve reliability (e.g. Bergman, 1985). Some examples are as follows:

— *Good design:* the choice of design and proficient engineering work are basic to high reliability.

— *Use of reliable components:* the reliability of the system as a whole depends in part on the reliability of components and subsystems.

— *Maintenance:* plant maintenance is of decisive importance in terms of reliability.

— *Continuous monitoring:* this applies to specific functions or features of the system. There are many ways of detecting functional deteriorations which are not yet so serious that they have given rise to defects which mean that the system fails to operate. For example, the level of vibration or temperature of bearings can be monitored, abnormal values providing an indication that failure may occur. The bearings can then be replaced before breakdown.

— *Regular testing* (see below)

— *Choice of safety margins* between load and strength (see below)

— *Redundant systems* (see below)

— *Fail-to-safe philosophy* (see below)

Regular testing

Regular testing of components and system functions can be an effective means of increasing reliability. This means that test routines are run at regular intervals to investigate whether different subsystems and important components are functioning. This has specially great importance for parts where hidden defects can arise and are only revealed when the function is specially required, e.g. in safety systems.

Choice of safety margins between load and strength

For strength to be sufficient, the design must be such that it can resist the load to which it is exposed without breaking. The load can take on various forms: the weight of objects, the pressure of water, the tension of electricity, etc. The concept of a safety margin is used to describe the difference between the size of the load and the resisting capacity of the system. Some systems may have a safety margin amounting to a factor of 10, while for other systems the margin is considerably narrower.

In reality, conditions are more complicated. Both load and strength can diverge from their theoretical values, and they should be regarded as variable quantities with a certain distribution. Strength can be reduced for a large number of reasons, such as corrosion, high temperatures in the surrounding environment, etc. A short discussion of how these issues are handled by Fault tree analysis is provided in section 8.4. Otherwise, the reader is referred to the specialized literature (e.g. O'Connor, 1991).

Redundant systems

Where greater reliability or safety is required, redundant functions can be introduced into a system. One such function is provided by a parallel piece of equipment (or routine), which is not needed for the system to operate, but which can take over if a defect arises. An example is a battery, that is permanently connected, which takes over in the case of the failure of normal power supplies.

Redundancy can be total, where, for example, there are two complete parallel systems. Or, it can be partial, where only the most important or most sensitive components are duplicated. The case of one lamp in a parallel electrical circuit is an example of partial redundancy

Cases of redundancy can be further classified into those which are active and those which are passive. Active redundancy describes situations where the extra system is permanently connected, such as the battery in the example above. Passive redundancy refers to cases where the reserve system is activated only when failure occurs, as, for example, when the battery is replaced by an engine-powered generator that is started when the mains voltage fails.

A further aspect of redundancy concerns whether a system is load-bearing or not. In the case of load-bearing redundancy, the extra component is permanently subject to load.

Fail-to-safe philosophy

It is not possible to completely avoid technical failures. One design philosophy that can be employed is to construct equipment so that it will always revert to a safe state if a failure occurs. With simple systems, this usually means just that the machine stops. More generally, the creation of such systems requires certain design principles and the selection of critical components which can only fail in a specific way.

Common cause failures

When calculating reliability or failure rates, it is often assumed that component failures are independent of one another. But, under some conditions, this assumption is false. One example is that of an error in the manufacture of components which results in a series of parts of one single design being defective in the same way. Several components can also be exposed to an unsuitable environment, such as a high temperature, leading to a deterioration in length of product life. Taking account of the possibility of common cause failure is particularly important in the case of redundant (parallel) systems where, in theory, a very high level of reliability can be be achieved. This, of course, is on the assumption that common cause failures will not occur.

Numerical examples

To illustrate quantitative aspects of different solutions, some simple numerical examples are provided below. Let us base these on one single module in a safety system. At a given point in time, the probability for failure is assumed to be 0.02.

To increase reliability, redundancy is achieved by coupling in an identical module in parallel. The probability for the simultaneous failure of both systems is 0.0004 (equation 4). Should a further module be connected in parallel, the probability for overall system failure would fall to 0.000008. Such a drastic reduction, however, is largely only of theoretical value, as it will probably be different types of common cause failures that are the main cause for concern.

Another way of increasing reliability is to reduce the time interval between tests. If tests are carried out ten times as frequently, the original probability of failure will be reduced to approximately 0.002. It is possible, however, that testing introduces new sources of failure, which will reduce the scale of the improvement.

Some comments

A solution involving redundant modules in a safety system increases the probability that it will function. At the same time, more components that might fail are introduced, which means that the probability increases that the production system will fail.

The system also becomes more complex, both to manufacture and to maintain, which means that costs rise. Higher reliability has a price.

Questions concerning techniques to achieve reliability are also taken up in the chapter on Fault tree analysis.

17

2.2 On human error

"To err is human", is an old proverb. It is nearly always the case that human error lies behind an accident. These errors can be of many different types: they can be simple, such as when someone hits their thumb; or more advanced cognitive errors, such as when an important safety system is disconnected. All people make unintentional mistakes, usually when they are performing the same action on many occasions every day. Usually, it is only when they have unfavourable consequences that they are detected.

The importance of human actions to the attainment of system safety has received increasing attention over a number of years, and there is now an extensive literature on the subject. The purpose of this section is to present a short discussion of the subject area. Otherwise, the technical literature and a number of reviews can be referred to (e.g. Petersen, 1982; Hale & Glendon, 1987; Reason, 1990).

The human factor

A popular reaction to an accident is to blame it on the "human factor". At least in Sweden, newspapers usually regard this as the main explanation. Often, it is a representative of an authority or a safety manager who couches the explanation in these terms. This is not wrong in itself. All accidents are related to human actions. People use a piece of equipment, and people also make decisions on how the equipment is designed, which material is used and how work is planned. *But, there is a certain "ring" to the term, the "human factor". It implies that accidents are due to irrational and unpredictable elements in a situation, and that nothing can be done about them.* Moreover, it is often the person who is injured who is regarded as the "factor". Such an attitude easily leads to passivity and is a barrier to accident prevention.

The reader should note that the expression, "the human factor", should not be confused with "human factors" in the context of ergonomics. In fact, the study of "human factors" is often considered to be equivalent in meaning to "ergonomics" and, in much of the modern literature, the term has acquired a very wide sense. One definition (Health and Safety Executive, 1989) runs as follows: "The term human factors is used here to cover a range of issues. These include the perceptual, mental and physical capabilities of people and the interactions of individuals with

18

their job and working environments, the influence of equipment and system design on human performance, and above all, the organizational characteristics which influence safety related behaviour at work."

On different types of human error

The incidence of human error varies considerably, and it differs between individuals. Moreover, the proneness of the individual to err varies with both time and situation. This can be due to a large number of different factors, both internal and external to the individual (e.g. Petersen, 1982).

There can be very different types of errors, and they can be defined and classified in different ways. One working definition has been suggested by Reason (1990). A slightly simplified version of his account is provided below.

Human error is "taken as a generic term to encompass all those occasions on which a planned sequence of mental or physical activities fails to achieve its intended outcome, and when these failures cannot be attributed to the intervention of some chance agency".

Slips and lapses are "errors which result from some failure in the execution and/or storage stage of an action sequence, regardless of whether or not the plan which guided them was adequate to achieve its objective".

Slips occur when an action does not go as planned, and they are potentially observable, e.g. slips of performance or slips of the tongue. The term lapse is often related to "more covert error forms, largely involving failures of memory, that do not necessarily have to manifest themselves in actual behaviour and may only be apparent to the person who experiences them".

Mistakes can be defined as deficiencies or failures in the process of making judgements or inferences. Mistakes are complex and less well understood than slips. This means that they generally constitute a greater danger, and they are also harder to detect (Reason, 1990).

On models and explanations

From the end of the 19th century and onwards, many have sought to understand why people make errors in their thinking and in the performance of actions. Reason (1990) has provided an interesting review of developments over the last hundred years.

The most renowned of the pioneers was Sigmund Freud (1914) who found meaning in what were apparently random and day-to-day slips and lapses. Analysis of the errors often permitted the detection of explanations in unconscious thought processes which had their origins in psychological conflicts.

In cognitive psychology, the idea of "schemata" plays a central role. The term was first adopted by Bartlett (1932). He proposed the view that schemata were unconscious mental structures composed of old knowledge, and that the long-term memory comprised active knowledge structures rather than passive experiences.

According to Reason (1990): "the current view of schemata is that they constitute the higher-order, generic cognitive structures that underlie all aspects of human knowledge and skill. Although their processing lies beyond the direct reach of awareness, their products — words, images, feelings and actions — are available to consciousness. The very rapid handling of information in human cognition is possible because the regularities of the world, as well as our routine dealings with them, have been represented internally as schemata. The price we pay is that perceptions, memories, thoughts and actions have a tendency to err in the direction of the familiar and the expected."

One model to which reference is often made is based on distinguishing between three different performance levels (Rasmussen & Jensen, 1974; Rasmussen, 1980).

1. On a **skill-based** level, people have routine tasks with which they are familiar and which are accomplished through actions that are fairly direct. The errors have the nature of slips or lapses.

2. On a **rule-based** level, people get to grips with problems with which they are fairly familiar. The solutions are based on rules of the type: IF/THEN. A typical type of error occurs when the person misjudges the situation and applies the rule incorrectly.

3. On a **knowledge-based** level, people find themselves in a new situation where the old rules do not apply. They have to find a solution using the knowledge that is available to them. On this level, errors are far more complex by nature, and may depend on incomplete or incorrect infor-

mation, or limited resources (in a number of different senses). Rasmussen has suggested that problem solution involves eight steps: activation, observation, identification, interpretation, evaluation, goal selection, procedure selection and activation. He does not state that each step is taken in this particular order. The decision maker can jump between steps in order to attain a solution to a problem.

Decision making can be regarded as a conscious and logical process during which costs and benefits are weighed against each other. Such an account presupposes that the alternatives are relatively clear and that a sufficient amount of definite information is available. In more complex situations, the limitations of people themselves mean that this is not a particularly accurate model. "The capacity of the human mind for formulating and solving complex problems is very small compared with the size of the problems whose solution is required for objectively rational behaviour in the real world — or even for a reasonable approximation of such objective rationality" (Simon, 1957). "The limitation in human information processing gives a tendency for people to settle for satisfactory rather than optimal courses of action" (Reason, 1990).

Violations

Violations represent a further type of human error. By a violation is meant an intentional action which is in breach of regulations, either written or oral. The intention, however, is not to damage the system. Deliberate intention to harm is better described as sabotage. It is difficult to draw a sharp dividing line between errors and violations, and perhaps this is not necessary. In some cases, conscious deviation from what is accepted practice may be seen as a deviation. In many situations, this also applies to risk taking. Violations can of course be committed both by people who work directly with a piece of equipment and those involved in planning and design.

There are many reasons why people act in breach of regulations. Some examples are given below:

1. The person does not know that the action constitutes a violation. He or she may not be aware of the regulation, or may not be conscious that the action in question represents an infringement.

2. The person is aware of the regulation, but forgets it, e.g. if it seldom applies.

3. The regulation is perceived as unimportant, either by the person himself or by those around him.

4. There is conflict between the regulation and other goals.

5. The regulation is thought to be wrong or inappropriate, either with or without reason.

Risk taking

On an individual level, and just as much so on an organizational level, accidents can be related to risk taking, i.e. actions are taken which are known to be dangerous and may also be forbidden. Risk taking has many similarities with committing a violation. It is unknown how great the problem of risk taking actually is. The accident frequency for men is higher than that for women, and, on average, young men run a greater risk of being injured than older men. Part of the explanation for this may lie in different propensities to take risks, but it is probably also the case that men who are young undertake tasks that are relatively more hazardous.

At the same time, it often pays to take a risk. Hazardous ways of working can be faster and less strenuous, and thereby give rise to higher productivity. The author's experience of the study of accidents is that over-ambition at work is a common explanation for the taking of risks. Even in risky jobs, accidents are relatively uncommon, happening to a person perhaps once every ten years. This is why it is generally the benefits of risk taking that are most visible, encouraging people to take risks and easily making risky behaviour habitual.

In general, piece work encourages risk taking. In forestry work in Sweden in 1975, a piece rate system was replaced by a more fixed form of remuneration. The frequency of accidents has fallen by around 30% (Sundström-Frisk, 1984). The taking down of already-felled but suspended trees is the job task where the benefits of risk taking are most apparent. For this task, the number of accidents fell by 70%.

Accident proneness

One early attempt to systematize knowledge on accidents involved focusing on the role of the individual. This resulted in a theory of accident proneness, devised by Farmer and Chambers (1926). They put forward the hypothesis that it was certain categories of people who were most likely to be the victims of accidents.

According to this theory, some individuals possess certain stable properties that make them particularly liable to accidents. A consequence of the theory is that accidents should be combatted by selecting individuals, e.g. by using various testing procedures, and allocating them tasks which are appropriate. However, it has proved difficult to distinguish properties of the individual from variations in exposure to hazards in the work environment. The theory does not provide a fruitful general model for explanation. In certain contexts, where stringent demands are placed on the individual to act safely, it has served to provide a basis for the selection of job tasks.

Why is it that more accidents do not occur?

The problem of accidents can also by approached from the opposite direction. From looking at a construction site, where people are climbing from place to place, where there is a lot of traffic, etc., it might seem that an accident must occur every day. The severe physical hazards do give rise to a higher frequency of accidents, but not one that is extremely high. One explanation lies in risk compensation. Construction workers adapt their behaviour in the light of the risks they face. In general, this question concerns safe and unsafe behaviour at work and how risks are perceived at the workplace (e.g. Hale & Glendon, 1987).

Human reliability

People do make errors, but more often they do things right. Instead of focusing attention on human error, an alternative starting point might be to regard people as a safety resource rather than a hazard. Even though this might appear to be a philosophical remark, this approach can provide a different basis on which systems can be designed.

An example

A man had worked for a long time with a packaging machine. After 10 years he received a severe squeeze injury from the machine. No special circumstances were found to apply and the accident was treated as the result of "inexplicable error"; perhaps, he had been tired and acted clumsily.

In his daily work, he had had to correct a disturbance to production roughly five times a day. This was dangerous if he made a mistake. On one occasion, he did make a mistake, and was injured as a result (after ten years, which is the average time between accidents per person). This means that he had managed to accomplish the task successfully 10 000 times without being injured. Should he not really be regarded as reliable?

At the packaging machine, the man corrected mistakes that were made elsewhere in the plant. One might ask about the defects that caused trouble so often. Why had these not been discovered and corrected? If there was no remedy, why were safer routines not used when he corrected the disturbance?

If human beings are regarded as the cause of occupational hazards, safety strategies might be based either on removing people from production through automation or on the strict supervision of work. A different approach would involve pointing to the role of the human being as a problem solver and safety factor in technical systems. Such an approach raises a number of questions, e.g. on the skills of operators and on the needs for information on the technical system, for organizational support and for the feeding-back of previous experiences.

Safer behaviour

Human error and risky behaviour are strongly affected by the technical design and organizational structure, and also by social patterns at the workplace. There are many different ways of getting individuals to behave more safely, which have a greater or lesser degree of success. Information campaigns to improve safety are quite common, but they tend to have effects that are both short-term and marginal (e.g. Saari, 1990).

Probably the most common way of attempting to promote safer behaviour is to introduce stricter rules, with supervision to ensure that they are followed. Attention is drawn to a type of erroneous behaviour, and an attempt is made to correct it. However, introducing rules which will achieve results is a difficult issue and requires a lot of thought (e.g. Hale, 1990).

There have been a number of experiments with "performance feedback", the goals of which are to get workers to make greater use of personal protective devices, employ safer working methods, etc. The feedback, for example, may consist of someone noting down the proportion of workers using a particular piece of safety equipment and reporting the results in the form of a diagram that is displayed where everyone can see it. Most studies have shown improvements in behaviour. On the other hand, the effect on accidents has been followed up more seldom (McAfee & Winn, 1989; Saari, 1990).

One pre-condition for either way of proceeding to succeed is that the technical and organizational conditions are right. Another is that correct and safe ways of working can be specified.

2.3 On systems and safety

The reasoning in this book is largely based on a systems perspective. In short, the approach can be characterized in terms of the view that a production system consists of a number of elements that must interact for a desired result to be achieved — and for accidents and other undesirable consequences to be avoided. The main elements are:

1. Technical equipment and physical conditions.

2. Individuals within the company.

3. Organization.

4. System environment.

The properties of the elements and the nature of the interaction between them determine the sizes of the accident risks. Attempts to reduce risk by treating just one of the four elements generally lead to rather ineffective solutions.

Another component of the systems approach to safety work involves taking account of the entire life cycle of the production system. Safety considerations must apply during planning, design, production start, operation and wind-up. The operational phase includes both normal and disrupted production, maintenance, system changes, etc.

The types of technical equipment that may be involved in accidents vary to a very considerable extent, ranging from hand-held knifes to computer systems which control production plants. Even with relative simple systems, designers have to adopt a large number of standpoints and make many different decisions. A number of these will concern safety issues, either directly or indirectly. These may be purely technical issues or might be concerned with how people use the machinery.

25

In the case of technical equipment, there has been a tradition that safety is relatively well regulated — by directives, norms and accepted good practice: construction materials must be appropriate and be sufficiently strong; dangerous substances shall be properly enclosed; protection against dangerous moving machine-parts shall be provided; safety guards and, sometimes, interlocking devices, etc., must be used.

Automated and computer-controlled equipment

One view that is sometimes put forward is that occupational accidents will become ever less of a problem as machinery comes to be more and more automated. As the human being has less contact with moving machine parts, the number of accidents will fall. Against this view, it has been pointed out that the number of accidents per employee is not falling in a number of specific industrial sectors (sect. 1.3) despite the steady automation process that has been in progress for many years. Moreover, the risks faced by large numbers of people seem to have risen as systems become controlled by computers and have ever more sophisticated functions. This can make it harder for the operator to anticipate what the machine will do. In many cases, safety functions can themselves cause problems, making supervision and maintenance more difficult. It appears that this is an area which is difficult to control by directive.

For this reason, the question of the safety of automated equipment is discussed quite extensively, thus also providing an illustration of system safety. In the first instance, automatic functions in different types of mechanical equipment are discussed. It can be supposed that in large chemicals processing plants, a great deal of care is taken in design and programming. Moreover, the functions (and problems) are rather complicated by nature.

One study of the different causes of accidents involving control systems (Backström & Harms-Ringdahl, 1984) found that a technical failure lay behind the accident in 19% of cases, while there was some type of operational deviation in 55%. The most common type of operator error was to implement start routines unintentionally (16%). The most common type of job task being implemented when accidents occurred was that of correcting a disturbance to operations (25%). In the case of accidents involving industrial robots, this proportion was considerably higher (60%). Thus, a strategy that was based on increasing technical reliability alone would have little effect on the number of accidents. *A further, more general, conclusion is that safety analysis which is wholly technically oriented is rather inappropriate for such systems.*

This particular study applied to accidents that took place in 1981. Since then the proportion of automatic control systems in industry has increased. A further study (Döös & Backström, 1990) treated accidents over the period, 1988-1990. The findings are in line with those of the earlier study: in 16% of cases, a technical defect had directly caused the accident; in 60%, an operator had been rectifying a disturbance to production.

These results draw our attention to a problem that Bainbridge (1987) has described as involving "the ironies of automation". It is often the case that designers regard operators as unreliable and inefficient, and therefore attempt to take automation as far as it can possibly go. One problem is that it is the errors of designers themselves which make a significant contribution to accidents. Another is that the designer who attempts to eliminate the operator will still leave to the operator the task of solving problems that cannot be automated sufficiently well. This applies, among other things, to the handling of disturbances to production which involve accident risks, a form of handling for which preparations are often poor and documentation inadequate (Backström & Harms-Ringdahl, 1986).

Little information on the size of the risks involving automatic equipment is available. The earlier study suggested that around 2% of occupational accidents in Sweden in 1981 might be accounted for in this way. But, in recent years, equipment that is either automatic or computer-controlled (to a greater or lesser extent) has been introduced at a rapid rate. Probably, the scale of the problem has increased, and will continue to do so.

A great deal of attention has been paid to accident risks involving industrial robots, both because of the apparent unpredictability and danger involved in their movements and probably also because there is something rather dramatic about machines that mimic the actions of human beings. Material in the Backström & Harms-Ringdahl study (1984) suggested an accident rate of 1 accident per 100 robot years while still lower frequencies were indicated in a review by Gossens (1991). The accident rate, therefore, may be regarded as relatively low. For example, a comparison can be made between an average industrial worker, who is exposed to one accident every tenth year, and an operator who supervises the operations of a number of robots. One interpretation is that

27

the high level of interest in robot safety has kept the number of accidents down.

One way of describing risks in a computer-controlled system based on the reliability approach has been adopted by Bell *et al.* (1983). The authors point to four causes of system error in a computer-controlled system, all of which can lead to a reduced level of safety. To these, a further three causes of a general nature have been added (Backström & Harms-Ringdahl, 1986):

1. Faulty or incomplete specification of safety requirements.

2. Random hardware failures.

3. Systematic hardware failures.

4. Software failures.

5. Errors introduced on system amendment, primarily programming errors.

6. Errors made when using the system, primarily the mistakes of operators, repairers, etc.

7. Organizational deficiencies, faulty or non-existent maintenance, a lack of manuals, etc. These do not constitute direct causes of system failure, but increase the probability of different types of errors.

Under each of these headings, there are a number of factors of which account should be taken. The greatest progress has been made in the field of random hardware errors, where reliability can be improved through redundant systems and fail-to-safe solutions for technical circuits.

One issue that has been discussed over a number of years is whether safety systems that are entirely controlled by computer software can be depended on. In nearly all the literature, such an approach is regarded as generally inappropriate, because sufficient safety is difficult both to achieve and verify. At the same time, the author's experience is that most computer-controlled machines have safety functions that are largely based on computer programs.

Tolerant and forgiving systems

One important property of technical systems concerns how important it is for the operator to always take the correct action. If a simple error directly gives rise to an acute hazard, the system should be regarded as dangerous. Systems can be "forgiving" to a greater or lesser extent. Such forgiveness might involve providing the operator with an indication that an error has been committed, before, for example, a machine movement is triggered off. Or, opportunities might be made available to correct an error, so that the error in itself does not have serious consequences.

The individual and safer ways of working

The conditions for the individual to work safely can be expressed in terms of three key words: **Know, Can** and **Will** (Bird & Loftus, 1976). Let us take the operator of an automated machine as an example. The safety of his work depends on a number of factors. Some of these are as follows:

— **That he knows how to work safely.**
   This depends on the training he has received for the job task in question. For example, machine manuals need to be designed so that they are comprehensible. Knowledge is needed on functional properties, both when the machine is in normal operation and when disturbances of different types occur.

— **That he can work safely.**
   A precondition for this is that it is easy to stop the machine, and that it can be re-started without having to go through a lengthy and cumbersome procedure. Information may have to be generated by the machine, so that the operator both understands how to act correctly and has what he needs to do so. Safety devices can be designed so that they are not a hindrance to doing the work, etc.

— **That he has the will to work safely.**
   The motivational condition is in many ways the hardest to satisfy, especially from a long-term perspective. But, if the operator is aware of the hazards, and if conditions that promote risk taking are kept to a minimum, the likelihood will be considerably greater that he will adopt a fairly safe way of working. It is also conceivable that this aspect of work should be given priority and/or supervised.

The organization

A fundamental aspect of safety concerns how machines are designed and maintained, how job procedures are planned and supervised, etc.

For this reason, it is necessary to carefully consider the role of company management. To this, there is a formal aspect, including the decisions taken by management, written regulations, delegation of responsibilities and task structure, etc. But there is also an informal one, as everything cannot be regulated and supervised.

Principles for decision making and problem solving in complex systems and different types of organizations are important determinants of how safe systems will be. A considerable amount of research has been conducted within this area (e.g. Rasmussen *et al.*, 1991).

Organizations have many variants, with respect to how they are set up and how they operate. These range from hierarchical systems, where decisions flow from top to bottom, through bureaucratic organizations, which are controlled by detailed regulations, to loosely-controlled organizations with diversified authority structures. An organization may have a fixed structure within which there are few opportunities for change, while another common approach is to "create one structure when the system is designed and then let the structure change during problem solving to take advantage of new opportunities" (Brehmer, 1991).

In some organizations, specific people are authorized to take certain decisions. In others, group decision making procedures are adopted, where an attempt is made to reach consensus among a number of people who are all capable of understanding the problem on their own. Distributed decision making is a more complex concept, involving the absence of a centralized control agent or decision maker (Brehmer, 1991).

Safety regulations

Safety regulations at a company can be used for specifying how a task should be undertaken, communicating information and controlling people's behaviour in hazardous situations. They can also be utilized to control higher-level functions that are of significance for safety, such as maintenance, planning, design, etc. The issues raised by safety regulations are also taken up in the chapters on Deviation analysis and safety management.

There is general agreement among most of the parties concerned that safety regulations are necessary and of great value. But, there are also a number of problems involved in drafting good regulations. One negative aspect is that regulations can be employed to compensate for poor design by placing greater demands on the person who is to use the equipment. Poor regulations, which are impossible to follow in practice, can also have the primary function of providing a scapegoat if an accident occurs.

Hale (1990) has provided an overview of the opportunities and limitations of safety regulations. One of his conclusions is that: "imposed safety rules are only needed where individuals' own rules are sufficient to prevent accidents. Company rules can be seen as a safety net and a support to cope with situations in which the person would otherwise be led astray by their normal behaviour."

Safety culture

The actual level of safety within a system is not governed only by technical design and directives. The individual's commitment, attitudes and perspective on risk are also important ingredients. A concept that has been employed to discuss these rather illusive factors is that of "Safety Culture". A recent report on operations within the nuclear power industry has taken up this concept and considered what it might imply in concrete terms (INSAG, 1991).

INSAG's definition is as follows:

"Safety Culture is that assembly of characteristics and attitudes in organizations and individuals which establishes that, as an overriding priority, [nuclear] plant safety issues receive the attention warranted by their significance."

Two general components of safety culture are identified by INSAG. The first is the framework which an organization must have, the second is the attitude of staff at all levels. Attention to safety consists of many elements:

— Individual awareness of the importance of safety.

— Knowledge and competence.

— Commitment, requiring demonstration at senior management level of the high priority of safety and adoption by individuals of the common goal of safety.

— Motivation through leadership, the setting of objectives and systems of rewards and sanctions, and through individuals' self-generated attitudes.

31

— Supervision, including audit and review practices, with readiness to respond to individual questioning attitudes.

— Responsibility, through formal assignment and description of duties and their understanding by individuals.

The commitment of management and individuals is one of the more central issues. It is also important not to adopt a rigid stance towards safety issues, but have an attitude which is open and questioning. INSAG provides a list of issues that might come to the fore if such an attitude is adopted:

— Do I understand the task?

— What are my responsibilities?

— How do they relate to safety?

— Do I have the necessary knowledge to proceed?

— What are the responsibilities of others?

— Are there any unusual circumstances?

— Do I need any assistance?

— What can go wrong?

— What could be the consequences of failure or error?

— What should be done to prevent failures?

— What do I do if a fault occurs?

Issues of safety have greater prominence within the nuclear power industry than in more usual forms of industry, where safety is given a wide range of priorities — from low to high. Some installations may have a low accident frequency despite serious physical hazards and lack of management interest in safety issues. In such cases, the carriers of the safety culture are the workers on the shop-floor, and the strength of this culture may be a function of traditions at the workplace or within the industrial sector.

2.4 System accidents

Simple explanations

Simple explanations of why an accident has occurred are common. There is a great desire to know "the cause". An accident is seen as the product of one event and is regarded as having one explanation. But, effective accident prevention requires investigation in-depth and an understanding of underlying factors.

Simple explanations, which treat just a selected portion of reality, may prevent problems from being solved effectively. For this reason, explanations or theories are needed, which will provide sufficient insight into why accidents occur and how they can be prevented. Even in practical work it is important to be aware of the explanations one is seeking.

A wide range of models

There is no uniform, universally-applicable theory of accidents or accident model. It has been suggested that the study of accidents should be regarded as an academic discipline of its own, designated by a term such as 'accidentology' (e.g. Andersson, 1991). Depending on the system with which one is concerned and on the purpose of the analysis, the theory and the methodology which is most appropriate may then be selected. For this reason, a variety of theories will be discussed at different places in this book.

There are a large number of different models available. There is a long tradition of research into individual behaviour and individual characteristics, and, on the medical side, there are a variety of different epidemiological models. By studying the interaction between the individual (the injury's "agent") and the environment, as has previously been done for infectious diseases, information on the causes can be obtained (Gordon, 1949). There is also a group of models which can be described as systems models. The emphasis in these lies on the functions of the system, human beings themselves being regarded as system components.

The Domino theory

The "Domino theory" was first launched by Heinrich (1931) in the 1930s, and has had great significance for practical safety work over many years. An accident is described in terms of a sequence of events, the most important element of the accident being either an unsafe act or a mechanical or physical hazard. If these elements can be eliminated, accidents can be prevented from occurring. However, the domino theory has also been subjected to criticism (e.g. Petersen, 1982). This tends to be based on the view that the theory describes accidents in far too simple a manner. Nor does it explain why unsafe actions are taken, or why mechanical or physical hazards arise.

Systems models

There are a number of theories which have been constructed from a perspective which is systems-oriented to a greater or lesser extent (e.g. Kjellén, 1984). Simply expressed, this means that there are always contributory causes for the occurrence of an accident. The point of departure consists in viewing the company as a system, with, for example, technical, human and organizational resources, which must interact for certain results to be obtained. On this view, an accident represents an abnormal system effect and might be due either to the failure of individual components of the system (including human beings) or a disruption to the interaction between them. Many of the various methods employed in safety analysis are derived, to a greater or lesser extent, from a systems perspective. This is especially true in the case of Deviation analysis and MORT. But, methods which have originally had just a technical orientation, such as Failure mode and effects analysis, can have a broader area of application.

Safety in systems

In simple systems with some type of risk source, an accident may occur if a person either makes a simple error or enters the danger zone. Similarly, a technical failure can easily lead to an accident. Examples include hitting in nails with a hammer or working with an unguarded press.

There are other systems with varying numbers of safety functions. To increase safety, guards can be installed to provide protection against dangerous machine parts. The guards can be equipped with an interlock which stops the machine if the safety guard is removed. There may also be safety regulations of different types, which govern the way in which work is conducted, etc.

Increasing the number of safety functions in principle reduces the probability that an accident will occur. As predicted by simple reliability theory, the number of independent failures necessary for its occurrence increases. The low probability of accidents that has been calculated for nuclear plants depends on the use of several independent safety systems, all with a high level of reliability. This is often called 'defence-in-depth'.

Accidents in any case

In systems with advanced safety systems, the theoretical probability of an accident is extremely low. In reality, accidents continue to occur.

Part of the explanation lies in that safety systems do not cover all eventualities. Perhaps, they are designed for certain normal situations and under certain assumptions. As an example, it might be mentioned that the most hazardous tasks arise in the course of correcting disturbances to automatic systems (see above). In systems which seem to offer an ambitious level of safety protection, the nature of such tasks may not have been sufficiently thought out.

A second part of the explanation lies in the fact that the systems do not function as planned, or that they deteriorate over time. It may be that certain components of a safety system are not required before a risk becomes acute. This may apply to both technical and organizational elements. Analyses of major accidents demonstrate that both types of problems arise (e.g. Reason, 1990).

3 Safety analysis

3.1 What is safety analysis?

Definitions

Safety analysis is defined here as a systematic approach to the identification and evaluation of factors that may lead to accidents. It also frequently includes the generation of proposals for increasing levels of safety. A further characteristic of the approach is that an attempt is made to obtain an overall picture of the hazards within a system. The section that follows contains a description of the various steps included in a safety analysis (on this definition).

The terminology used in safety analysis tends to be rather variable, which is related, among other things, to the fact that analyses of risks are conducted within a variety of professional areas, and in various ways. This means that the meanings of a number of concepts also vary. The terms, risk analysis (or assessment) and hazard assessment are used in a rather similar way to that of safety analysis.

Safety analyses can provide quantitative measures of certain types of risks. This may mean that numerical values are generated for the likelihood of certain events occurring, or that calculations are made of possible consequences, or that weighted measures are constructed, e.g. according to the principle of probability times consequence. To emphasize that an analysis is quantitative, the terms Probabilistic risk analysis (PRA) or Probabilistic safety analysis (PSA) are frequently used. Qualitative analyses do not generate such numerical values. This book is generally concerned with qualitative analyses, although some quantitative aspects are discussed.

A Nordic cooperative venture (SCRATCH, 1984) which focused on safety analyses proposed a number of definitions. Safety analysis is described as a systematic way of investigating or analyzing risks, sources of risk (hazards) or accidents. Risk analysis is defined as a systematic form of analysis, with the specific purpose of determining the risk that is related to a system, i.e. it is also a form of quantitative analysis.

This book is concerned with analyses designed to prevent "common" accidents in "normal" installations, a subject area in which safety analysis and risk analysis have not become particularly well established and where interpretations of what an analysis should consist may vary. For this reason, a piece of advice to anyone about to conduct a safety analysis is to be careful to define clearly the aims of the analysis and what it involves.

The systematic approach

Let us suppose that a particular production system is to be analyzed. The analysis might apply to an existing installation or to production facilities that are still at the planning stage. There are several different aspects to a systematic approach:

— Gathering of information on the system provides the basis for the analysis and must be carried out systematically.

— The entire system and the activities within it should be included in the analysis. The analysis needs to be designed systematically so that important elements are not overlooked. A main thread must be identified and followed.

— A systematic methodology is required for the identification of hazards.

— The risks to which these hazards give rise need to be assessed in a consistent manner.

— A systematic approach is required even when safety proposals are to be generated and evaluated.

The adoption of any one specific method might mean that certain types of hazards will be observed, while others may not be detected. If, for example, a method which focuses on technical failures is adopted, only certain types of conceivable accidents may be covered. By using supplementary methods, better coverage of the system and different types of hazards is obtained.

The systematization of experience

A method for safety analysis can also be seen as a compressed account of previous experiences. For example, the check lists that are used in several of the methods represent summaries of what has previously been found to be important in terms of the identification of hazards.

The development of analyses and safety activities are very much "accident-driven". Or, as Reason (1990) puts it, "events drive fashions". People have been forced to rethink, in one way or another, by their own experiences. Perspectives on accidents and strategies for the analysis of risks have been governed to a considerable extent by accidents that have already occurred.

3.2 Components of a safety analysis

A safety analysis is not a single activity, but consists of a number of co-ordinated steps which jointly make up a procedure. Figure 3.1 provides one example of a safety analysis procedure. A consistent theme of this book is that we assume that the aim of an analysis is to achieve a reduction in the level of risk. For this reason, decisions on and the implementation of safety measures have been included. It should be remembered that there are a variety of other flow charts, used by various authors, and which have different areas of application.

The three central elements in the figure are the identification of hazards, the assessment of risks, and the making of proposals for safety measures. The form that these activities take is related to the method employed, while the other elements are of a more general nature.

PLANNING

The planning of analyses is more extensively discussed in chapter 12. One of the first steps is to take the decision to conduct a safety analysis. This involves consideration of:

— What is to be analyzed, which limits to the analysis are to be set and what assumptions are to be made.

— The aim of the analysis. Is it designed to generate a proposal to increase the level of safety? Or, is it a general evaluation of the extent to which plant or equipment is hazardous. In the latter case, the stage "Proposals for safety measures" disappears from the analysis.

— The scale of the analysis, the allocation of resources, etc.

— The choice of methods and the manner of approach.

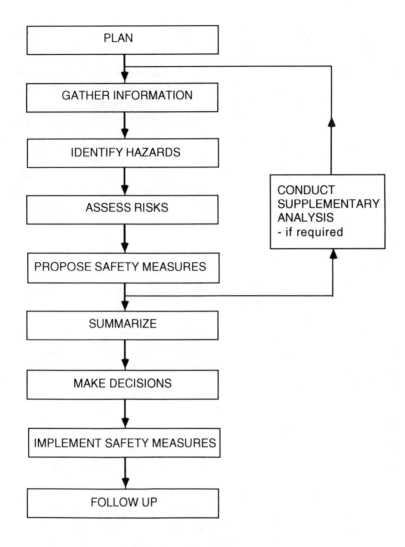

Figure 3.1 Stages of safety analysis procedure, and how results are used

GATHERING INFORMATION

It is information that provides the base from which an analysis can proceed. Information on the system to be analyzed is needed. This applies to its technical design, how the system functions and which activities are undertaken. To a great extent, the need for information is governed by the choice of methods to be employed.

Various types of information on the problem can be utilized. These may concern accidents that have occurred, near-accidents or disturbances to production. If probabilistic analyses are to be conducted, data on frequencies of failure for the components used in the system are also needed.

In the cases of analyses of installations that have been in operation for some time, this information is relatively easily accessible. On the other hand, where the analysis concerns production facilities that are still at the planning stage, it is more difficult. Information can be obtained from drawings, written and oral descriptions, and from experiences of similar installations.

IDENTIFYING HAZARDS

The central component of most safety analyses is the identification of hazards and other factors in the system that might lead to accidents. One aim should be to to discover the major sources of danger and which factors might trigger off an accident. The method selected determines how the process of hazard identification proceeds. When a specialized method is used, certain types of hazards are discovered, but others may be overlooked.

ASSESSING RISKS

An assessment of the risks posed by the identified hazards needs to be made. In a quantitative analysis, estimated values of probabilities and consequences may be obtained. On the basis of these values, an evaluation is then made as to whether the risk is acceptable or safety measures are necessary. In the case of qualitative analyses, as discussed in chapter 12, an evaluation is made directly. Sometimes, directives and norms are available, but not all of these are useful for practical assessment.

PROPOSING SAFETY MEASURES

In general, the risks can be reduced through one or several safety measures. The reduction can apply to either the consequences or the probability that they will occur. Some of the methods include a systematic procedure for the identification of potential safety measures.

SUPPLEMENTING THE ANALYSIS

During the planning of the analysis, only an incomplete picture of the hazards in the system is available. In the course of conducting the analysis, it might be discovered that more detailed examinations are required, or that a supplementary method is appropriate.

SUMMARIZING

The results of the analysis are summarized so as to provide a basis for decision making. The summary might include a list of the hazards observed, proposals for safety measures, and an account of the assumptions and conditions under which the analysis was conducted.

MAKING DECISIONS

We assume that the summary is then used as a basis for decision making. An analysis that is not put before the right decision making forum is not of much value.

IMPLEMENTING SAFETY MEASURES

For the analysis to have an effect, the safety measures decided upon must, of course, be implemented.

FOLLOWING UP THE ANALYSIS

It is also a good idea to make plans to follow up the analysis. This can involve making certain checks on the analysis, establishing that measures as decided have been implemented, and examining results at a later date. For example: Has the number of accidents fallen? How has production been affected?

3.3 A short methodological overview

There are many different methods of safety analysis (for further discussion, see chapt. 10). Clearly, anyone who is to work with safety analysis must make a choice between the large number of methods available. In general, it can be said that any one specific method will only cover a limited part of the risk panorama.

Some of the criteria that might be used in the choice of method are as follows:

— That the method provides the support necessary to sustain a systematic approach (as discussed above in sect. 3.1).

— That the method is easy to understand and apply.

— That an analysis can be conducted even when information on the system is incomplete. For example, this may be the case when an analysis is conducted of plant or equipment that is still at the planning stage. This may give rise to poorer accuracy, but the analysis is still worthwhile.

— That an analysis can be conducted with a reasonable amount of effort, taking anything from part of a day to one or several weeks.

This book contains the author's own selection of methods. The selection has been guided by several factors, an important one being the types of systems which are to be analyzed. It is also the case that no method will automatically guarantee good results. The methods should be regarded as an aid to creative and well-structured thought.

Table 3.1 provides a sample of the methods presented in this book. A more extensive overview of a number of methods and a comparison between them are presented in chapters 9 and 10. The methods are classified as follows:

— Analysis from a technical perspective.

— Man-machine interaction.

— Analysis of the organization.

Model of the system
Conducting an analysis requires the simplification of a complex reality, the construction of a model. The three categories above reflect, at least to a certain extent, the types of models on which the methods are based.

Model of the accident
The methods also include a more-or-less explicit explanation of how accidents occur. These can also be seen as models or theories.

Table 3.1 *Some methods of safety analysis*

Method	Comments
Technically-oriented	
Energy analysis	Identifies hazardous forms of energy (chapt. 4).
HAZOP	Analysis of deviations in chemicals processing plants (chapt. 7).
Fault tree analysis	Logical representation of the causes of a specific event, usually an accident (chapt. 8).
Man-machine interaction	
Job safety analysis	Hazards in relation to the implementation of work tasks (chapt. 5).
Deviation analysis	Hazards caused by deviations in the functioning of equipment, human beings and the organization (chapt. 6).
Organizational analysis	
MORT	Organizational factors of importance to safety (chapt. 9).

The analytical procedure

For four of the analyses mentioned above, the procedure for analysis represents an important part of the method. This means that the different steps are taken in accordance with a specific plan (or schema). This facilitates undertaking the analysis, and also makes it easier to plan. These four methods are Energy analysis, HAZOP, Job safety analysis and Deviation analysis.

Some of the important steps that these methods have in common are as follows:

1. A system is divided into several components, which involves the construction of a simplified model of the system. This step is called structuring.

2. For each component of the system, sources of risk (hazards) or other factors related to the risk of accidents are identified.

3. Some form of risk assessment is carried out.

4. In most cases, a stage at which safety measures are proposed is included.

Quick analyses

To obtain a quick overview of hazards at a plant or from a piece of equipment, some type of rough analysis can be conducted. Such an analysis represents a compromise between thorough analysis and less systematic observation (chapt. 9).

4. Energy analysis

4.1 Principles

The idea that lies behind Energy analysis is a simple one. For an injury to occur, a person must be exposed to an injurious influence — a form of "energy". This may be a moving machine part, electrical voltage, etc.

In using this method, the concept of energy is treated in a wide sense. Energy is something that can damage a person physically or chemically in connection with an event. An injury occurs when a person's body is exposed to an energy that exceeds the injury threshold of the body. The purpose of the method is to obtain an overview of all the energies in an installation which can cause acute injuries to a human being.

This line of approach, of seeing the reason for injuries in terms of energy, was first developed by Gibson (1961) and Haddon (1963). It has been shown to be fruitful, and has been further developed and discussed in several books and reports (e.g. Hammer, 1972; Haddon, 1980; Johnson, 1980). An additional feature of the account presented here is the way in which the analytical procedure is broken down into a number of defined steps (Harms-Ringdahl, 1982).

Thinking in energy terms is based on a model of systems (and of reality) that contains three main components:

1. That which can be harmed, usually a human being but possibly an object, piece of equipment or industrial plant.

2. Energies, which can cause harm.

3. Barriers, which prevent harm from being caused, such as safety guards for machinery.

In the model that applies to accidents, a person or object comes into contact with a harmful energy. This means that the barriers have not provided sufficient protection.

Harmful energy can take on many forms, such as an object 'at a height (from which it may fall), electrical voltage, etc., i.e. energies in a traditional sense. By adding poisonous and corrosive substances, etc., a fairly comprehensive picture of the acute injuries that might affect a human being is obtained.

45

4.2 Energy analysis procedure

An Energy analysis contains four main stages, plus making preparations and concluding the analysis. It is usually best to fully complete each stage before going over to the next. As an aid to the analysis, a specially-designed record sheet can be used (example in table 4.3).

Preparing

Before embarking on the analysis itself, a certain amount of preparation is required. This concerns obtaining information on the "object" of study, and setting boundaries between it and other parts of the installation (see also the discussion in chapt. 12).

Analysis

The Energy analysis itself is divided into four main stages.

1. STRUCTURING

The purpose of the structuring stage of the analysis is to divide the

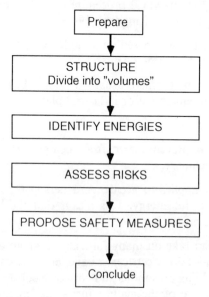

Figure 4.1 *Main stages of procedure in Energy analysis*

system into suitable parts, which are then analyzed one at a time during the stages of the analysis that follow.

In general, a classification is made in accordance with the physical characteristics of the installation under study. In principle, the plant or equipment is divided up into "volumes" (spatial segments) — "volume structuring". If the analysis is applied to a long production line, it is appropriate to go from one end of the line to the other. If this does not cover the entire area to be analyzed, supplementary volumes are needed. It can be imagined that the installation is divided up into "boxes". If there is a natural central point, the installation can be split into concentric circular segments.

This means that the boundaries of the entire system to be analyzed should also be thought of in volume terms. After provisional structuring, a check should be made as to whether any component has been omitted or "lost" in some way. Sometimes, it can be wise to add an extra "volume" to cover anything which lies outside the area where the object in question is located.

2. IDENTIFYING ENERGIES

For each "volume" (or "box"), sources of energy are identified. The check list of energies shown in table 4.1 can be used as an aid for this.

One problem is to determine the lowest level of energy with which the analysis should be concerned. There is a trade-off between comprehensive coverage of hazards and the avoidance of trivia. This decision should be made in the light of the aims and level of ambition of the application in question. However, an energy should not be excluded just because it seems unlikely that a human being will be exposed to it.

3. ASSESSING IDENTIFIED ENERGIES

Each identified form of energy is assessed. In the first instance, the assessment concerns conceivable injuries to which a human being might be exposed. One difficulty is that, in terms of injury, any one energy may have a variety of consequences.

The assessment also takes account of any relevant barriers and the likelihood that injuries will occur. One example (discussed further in chapt. 11) of a simple composite (weighted) scale is as follows:

0. No danger.

I. Acceptable risk, no safety measure required.

II. Certain degree of risk, safety measure required.

III. Serious risk, safety measure essential.

4. PROPOSING SAFETY MEASURES

When all energies have been assessed, an attempt is made to arrive at safety measures. For energies which are important, efforts are made to apply the methodology that is expressed in table 4.2.

At the beginning, it is a matter of getting ideas. It is good to be able to suggest a variety of solutions, as it is not certain that the first will be the most effective. Then, the solutions with which one wants to go further can be selected.

Concluding

The analysis is concluded by preparing a summary, listing, for example, the most important energies and proposals for safety measures.

Energy check list

Table 4.1 shows a check list for different types of energies which is designed for use as an aid to identification. For most categories, the link between energy and injury is obvious. But, some types of energies may require further comment.

"Chemical influence" is treated here as an energy that may give rise to injury. In some cases, it is possible to conceive of this influence in terms of the chemical having a micro level effect on human cells. "Asphyxiating" chemicals are gases or liquids which are not poisonous in themselves, but which restrict or eliminate access to air. This subcategory might refer to the possibility of being exposed to a suffocating gas or of drowning in water.

The final category on the check list has the heading "Miscellaneous". It is included to provide an additional check on whether the focus of the analysis has been too narrow in a technical sense. An extra check is ob-

Table 4.1 *Check list for Energy analysis*

1. POTENTIAL ENERGY Person at a height Object at a height Collapsing structure Handling, lifting, etc.	6. HEAT & COLD Hot or cold object Liquid or molten substance Steam or gas Chemical reaction
2. KINETIC ENERGY Moving machine part Flying object, spray, etc. Handled material Vehicle	7. FIRE & EXPLOSION Flammable substance Explosive: - material - steam or gas - dust Chemical reaction
3. ROTATIONAL MOVEMENT Machine part Power transmission Roller/cylinder	8. CHEMICAL INFLUENCE Poisonous Corrosive Asphyxiating Contagious
4. STORED PRESSURE Gas Steam Liquid Coiled spring Material under tension	9. RADIATION Acoustic Electromagnetic Light, inc. infra and ultra Ionized
5. ELECTRIC Voltage Condenser Battery Current (inductive storage and heating) Magnetic field	10. MISCELLANEOUS Human movement Sharp edge Danger point, e.g. between rotating rollers

tained on whether the movements of a human being might involve the risk of falling, stumbling, colliding with protruding objects, etc. The "Sharp edge" and "Danger point" subcategories do not really refer to forms of energy, but such items can be seen in terms of energy concentrations when a person or piece of equipment is in motion.

Table 4.2 *Finding safety measures using Energy analysis*

Safety measure	Examples
The energy 1. Eliminate the energy	Work on the ground, instead of at a height. Lower the conveyor belt to ground level. Remove the hazardous chemicals involved in the process.
2. Restrict the magnitude of the energy	Lighter objects to be handled. Small containers for dangerous substances. Reduce the speed.
3. Safer alternative solution	Less dangerous chemicals. Handling equipment for lifting. Equipment requiring less maintenance.
4. Prevent the build-up of an extreme magnitude of energy	Control equipment. Facilities for monitoring limit positions. Pressure relief valve.
5. Prevent the release of energy	Container of sufficient strength. Safety railing on elevated platforms.
6. Controlled reduction of energy	Safety valve. Bleed-off. Brake on rotating cylinders.
Separation 7. Separate the human being from the energy flow a) in space	One-way traffic. Separate off pedestrian and motorized traffic. Partition off dangerous areas.
b) in time	Schedule hazardous activities outside normal working hours.
8. Safety protection on the object (the energy source)	Machine safety guards. Electrical insulation. Heat insulation.
The human being 9. Personal protective equipment	Protective shoes, helmets.
10. Limit the consequences when an accident occurs	Facilities for stopping the energy flow. Emergency stop. Emergency shower facilities. Specialized equipment for freeing a person (if stuck).

The list contains a few deliberate inconsistencies. Some categories do not refer to energies in a physical sense, but are conceived of as means of getting at sources of risk (hazards). This applies, for example, to "Collapsing structure". The object to be studied might be a heavy installation, such as a liquor tank. The energy in question is the potential energy of the tank. The category "Collapsing structure" draws attention to major pieces of equipment and the possibility that they can overturn or have other defects. Similarly, the subcategory "Handling, lifting, etc." is used to cover the potential and kinetic energy of a handled object. The idea is that problems related to the manual handling of materials can also be treated at the identification stage.

A systematic approach to safety measures

One of the advantages of Energy analysis is that it provides a systematic means for developing safety measures (e.g. Haddon, 1980; Johnson, 1980). Table 4.2 outlines the approach and provides examples of measures designed to reduce risks that are generated by adopting the energy approach (see also p. 54).

4.3 Example

In this example, a tank for the storage of sodium hydroxide (lye) is to be acquired. There is a desire to make a preliminary assessment of the hazards involved. A starting-point is a sketch of the installation, and a simple Energy analysis is made on the basis of this.

System description

Concentrated sodium hydroxide is to be stored in a stainless steel tank. At lower temperatures the lye is viscous, and heating equipment using an electrical current is needed. The tank is filled using a tube equipped with a valve. On the top of the tank, there is a manhole and a breather pipe. Under the tank, there is a pit. A ladder has been permanently installed to provide access to the tank. Not visible on the sketch is a tube with a tap, used to tap off the liquid.

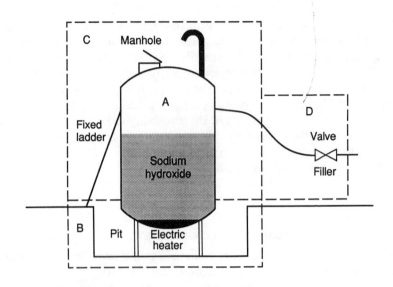

Figure 4.2 Liquor tank

Preparing

Our starting material consists of the sketch and the description above. The limits of the system are set by what is visible on the sketch. Table 4.3 shows an example of the record sheet used.

1. STRUCTURING

A division is made into four volumes, as shown in figure 4.2:

A. The tank.

B. The pit (the space under the tank).

C. The area surrounding the tank.

D. The filler tube and its surroundings.

2. IDENTIFYING ENERGIES

Let us start with the tank (Volume A) and follow the check list (table 4.1).

First, there is Potential energy (1).

- "Person at a height" will be relevant when someone goes down into the tank for servicing.

- The level of the liquid is above that of the tapping-off tube. If the valve is opened or if a connecting tube fails, the lye will run out.

- The tank has great mass. It requires stable supports ("Collapsing structure").

Then, Stored pressure (4) may be relevant. If the ventilation system fails, there will be high pressure on re-filling, or low pressure on tapping-off. (The pressure of liquid was earlier treated as a form of potential energy.)

Electric (5) refers to the power supply to the heating element. Insulation failure is hazardous. Lye is electrically conductive.

Heat & cold (6) applies to the heating element. Over-heating could occur if the liquid level is low, or the supply of energy too great.

Chemical influence (8) is obviously relevant because lye is highly corrosive. It is certainly the most obvious and the greatest hazard in the system.

The Miscellaneous (10) category provides an opportunity for a variety of items to be be taken up. It is not necessary to think strictly along energy lines. For example, one might wonder about the manhole. It has to be large enough, and there must also be space for a ladder.

Then we continue with the pit (Volume B) and continue through the remaining volumes.

3. ASSESSING THE RISKS

The risks are assessed, and the scale 0 – III (above) is used.

(The assessment in this example reflects the thoughts of the imaginary study team.)

4. PROPOSING SAFETY MEASURES

Before making a start on conceiving safety measures, the possibility of grouping the hazards into more general categories should be investigated. Lye comes up throughout the analysis, and all cases where it occurs might be considered simultaneously. We imagine that a number of technical safety measures will be conceived and proposals for job routines made. These will apply to lye in general. In the table, the term "lye package" is used to stand for these safety measures. In addition, a check should be made on each place where lye is noted on the record sheet to see if extra control measures should be taken.

It can be difficult to make concrete safety proposals for some hazards. In such cases, it can then be stated, for example, that job routines must be established or that a further investigation needs to be made. Or, it is even possible to note that no solution has been thought of, but that the problem still requires attention.

Let us take the hazards created by the lye itself as an example of how a systematic approach to the consideration of safety measures can be employed.

1. Eliminate.	Can lye be removed from the production process?
2. Restrict.	A smaller tank?
3. Safer alternative solution.	Can the lye be replaced by another chemical, or can a more diluted mixture be used?
4. Prevent build-up.	Safety device to prevent over-filling?
5. Prevent release.	Secure connection for hose on filling, and a method for emptying the filler tube after the tank has been topped up.
6. Control reduction.	Overflow facilities in case of over-filling.
7. Separate off the human being.	Prohibit unauthorized entrance and fence off the area.
8. Safety protection on the object.	Keep the filler tubes in a locked cupboard.
9. Personal protective equipment.	Protective clothing.
10. Limit the consequences.	Emergency shower facilities. Water for flushing. Draining facilities. Emergency alarm. "First aid" facilities. Make the pit under the tank sufficiently large.

Table 4.3 *Record sheet from the Energy analysis of a liquor tank*

Volume / Part	Energy	Hazard / Comments	Evaluation	Proposed measures
A Liquor tank	Person at a height (4 metres)	Falls down / During service	II	Routines for service
	Level of lye	Contact with lye /Lye can run out	III	"Lye package"
	Weight of tank (10 tons)	Falls or collapses / If damaged or poor design	I	— (Standard construction)
	Excessive pressure	- / On filling	O	— (Good ventilation)
/Heating system	Electric (380 volts)	Shock / If poor insulation, lye is conductive	II	Check proposed installation
	Heat	Burning / Temperature is low	I	
B Pit	Height (1.4 metres)	Falls	III	Railings and fixed ladder
/Heating system	See above	—	—	
C Outside tank	Person on platform or ladder	Falls	III	Suitable design of platform, railings and fixed ladder
	Tools and equipment at a height	Fall on people below	II	As above
D Filler tube	Level of lye	Contact with lye remaining in tube	III	"Lye package", routines, and lock filler tubes in safe place
General	Lye (10 tons)	Corrosive	III	"Lye package"

55

Concluding

After completing the analysis, a summary is prepared. In this case, it will consist of a list with guidelines to be applied during continuing design and planning.

Remark

The results of the analysis are not remarkable, but they provide a more complete picture than otherwise would have been available. In this example, it would have been possible to stare blindly just at the hazards created by the lye itself. Then, probably, only some of the problems would have come to light, and only some conceivable safety measures been suggested. A more extensive analysis would have dealt with situations that arise in the course of re-filling the tank, etc. Therefore, in this case, a supplementary method should be employed.

4.4 Comments

A simple method

The method can seem cumbersome, involving long check lists for each stage of the analysis. But, with a little experience, it is simple and quick to use. The identification stage can be completed in just one or a few hours, even with quite large systems.

Examples of pitfalls and problems when using Energy analysis:

— Some "volumes" are missed.

— Too much time is spent on details, e.g. trivial energies.

Identifying energies

Sometimes, judgement is needed in weighing up which energies should be included in the analysis. In principle, anything that can lead to an injury to a human being should be included. That contact with an energy is unlikely is no reason for it not to be included.

On the record analysis sheet, a note should be made of which specific energy is concerned. Simply repeating the names of categories on the check list should be avoided. For example, "lye" and not just "corrosive substance" should be entered, while "tool at a height" should be entered instead of "object at a height".

Assessing risk

The most practical way of proceeding is to identify energies in all volumes before embarking on the next stage of the analysis. This permits a better overall picture to be obtained and a more consistent form of risk assessment to be applied.

In the example of the liquor tank, the analysis was based on the idea of assessing whether the risk was acceptable or not. In Energy analysis, a good alternative is to assess hazards using a scale of measures of conceivable injury, e.g. whether the magnitude of the energy would lead to fatal, serious, minor or trivial injury (for further discussion, see chapt. 11).

Ideas for safety measures

In order to generate ideas for safety measures, it is best to think freely and try to come up with as much as possible. The check list is designed to be an aid to the imagination and a means for getting away from rigid lines of thinking. It is meant to provide different angles of approach. When a body of ideas has been built up, then the process of sifting and processing the ideas can begin.

Energy magnitude

In many cases it is possible to specify the magnitude of the energy with which one is concerned, e.g. in terms of its height, weight, or speed. This provides a more concrete basis for the assessment of risk.

With respect to rotating objects, this can be difficult. Then, however, it is possible to convert rotational energy into one of its equivalent forms. What would it mean in terms of velocity if the object came loose? To which height could the object be lifted? In principle, the moment of inertia could be calculated, and, on the basis of this and the rotational energy, the magnitude of the energy could be derived.

In most cases, however, a simple calculation, which has the advantage of being easy to remember, can be made. If the rotating object has the

57

bulk of its mass on its periphery, e.g. a tube that rotates along its longitudinal axis, the magnitude of the energy is obtained from the familiar expression:

$$W = m\ v^2/2$$

Where W stands for energy (in joules), m is the mass (kg), and v is the circumferential velocity in metres per second. The equivalent height h is:

$$h = v^2/2g$$

Where g stands for acceleration due to gravity, usually 9.8 m/s^2.

In the case of a solid cylindrical object, the moment of inertia is half that of a hollow cylinder. The following expressions are then applicable:

$$W = m\ v^2/4,\ \text{and}$$

$$h = v^2/4g$$

Let us take the example of a paper rolling machine. Paper is wound onto a roller, and a roll can weigh up to several tens of tons. Paper may be rolled at a speed of 2 000 metres/minute. The equivalent potential energy is that of a body of mass at a height of 28 metres.

5 Job safety analysis

5.1 Principles

In Job safety analysis, attention is concentrated on the job tasks performed by a person or group. The method is most appropriate where tasks are fairly well defined. The analysis is based on a list of the phases into which a job task can be broken down. The approach consists of going through the list point by point and attempting to identify different hazards at each phase. Sometimes, this procedure is called Job safety analysis, sometimes Work safety analysis.

The method is not based on any explicit model of how accidents occur. The production system is seen from the perspective of either the worker or the job supervisor. It is divided up into tasks controlled by machines and those governed by job instructions. However, the picture of accidents is fairly close to the energy model, and some job safety manuals contain check lists of different energies.

One of the advantages of the method is that it is straight-forward and relatively easy to apply. Several descriptions of the method have been published (e.g. Grimaldi, 1947; McElroy, 1974; Heinrich *et al.*, 1980; Suokas & Rouhiainen, 1984).

5.2 Job safety analysis procedure

Figure 5.1 shows that Job safety analysis consists of four main stages, each prior stage having to be completed before the next can be embarked upon. Preparation includes defining and setting the boundaries of the job tasks that are to be analyzed, and gathering information on instructions where these are especially important for the implementation of a task. It is also of benefit to involve a team of people in the analysis, preferably including a person who does the job and a job supervisor. For the analysis, a special record sheet is used. An example is shown in table 5.1.

1. STRUCTURING — the list of job tasks
A suitably detailed list of the different phases of the work under study is

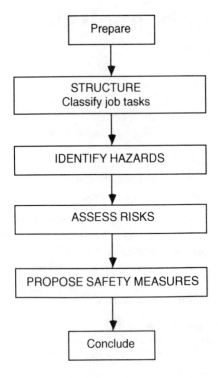

Figure 5.1 Main stages of procedure in Job safety analysis

prepared. Good basic material consists of standard job instructions, but these should not be assumed to be either complete or correct. It is also important to take account of unusual tasks and those that are only seldom undertaken. The following items should be considered:

— The standard job procedure.

— Preparations for and finishing off the work.

— Peripheral and occasional activities, such as obtaining materials, cleaning, etc.

— Correcting the disturbances to production that may arise.

Depending on the type of work, the following two components may also be included:

— Maintenance and inspection.

— The most important types of repairs.

2. IDENTIFYING HAZARDS

The subtasks on the list are gone through one at a time. A number of questions are posed in relation to each of these:

— Which types of injuries can occur?

  — Pinch/squeeze injuries or blows, moving machine parts, objects in motion or at a height, etc.

  — Cuts or pricks/stabs, sharp objects, etc.

  — Falls, working at a height, etc.

— Can special problems or deviations arise in the course of the work?

— Is the job task difficult or uncomfortable?

— Can other ways of doing the work arise?

It is not necessary to limit the analysis to accidents. Contact with chemicals, ergonomic problems, etc. may also be included.

3. ASSESSING THE RISKS

Each identified hazard or problem is assessed. A variety of grounds for classification and risk assessment may be utilized (for a further discussion, see chapt. 11).

4. PROPOSING SAFETY MEASURES

The next stage of the analysis is based on the hazards that are regarded as serious. When going through the record sheet, an attempt is made to propose ways of reducing the risks. The measures can apply to:

— Equipment and task aids.

— Work routines and methods (Can the work be carried out in a different way?).

— Elimination of the need for a certain job task.

— Improvements to job instructions, training, etc.

— Planning how to handle difficult situations.

— Safety devices fixed to pieces of equipment.

— Personal protective equipment.

This stage of the analysis principally concerns the creation of ideas. It is of benefit if ideas for several alternative solutions are generated. Several measures may be required to reduce a given risk. A particular safety measure may be hard to implement so an alternative might be needed.

Several different items that are similar in one way or another may be merged into one, e.g. if the hazards have similar causes or if a common safety measure is required. The proposed measures are entered on the record sheet.

Concluding

The analysis is concluded with a summary of the results. In simple cases, the record sheet itself may be used to report the results. The list of job tasks and the record of the analysis may also be used, fairly directly, to produce an improved set of job instructions.

5.3 Example

Job safety analysis has been employed in some of the examples presented later in this book (chapt. 15). Table 5.1 shows an extract from the record sheet of an analysis of a machine for the rolling and cutting of paper (from sect. 15.2).

One common application of the analytical method is in job planning. A supervisor may consider a repair that has to be made. He goes through his list of what is to be done together with the repairers. This enables them to identify hazards at various work phases and determine which safety measures are needed.

Such an analysis is informal, and the records of the analysis, etc. are not so important. A decision can be made immediately, and people appointed to take responsibility. The extra time required for the analysis may be about an hour. The quality of the analysis may not be so high, but the reduction in risks can still be significant.

Table 5.1 *Extract from the record of a Job safety analysis applied to a paper rolling machine (from sect. 15.2)*

Job task / Part	Hazard	Comments	Evalu-ation	Proposed measures
Removal of old base roll	Roll falls down	Rather heavy (40 kg). The roll can get stuck, or the operator can loose grip	II	Lifting equipment and adjustable holding facilities
Taking away packing band on new roll	Cutting injury	Sharp steel band Packing band recoils, due to tension Erroneous method (knife)	II	Use of proper tools Include this task in instruction manual
Installation of new base roll	Roll falls down	Heavy (2 tons). Mistake in operating lifting equipment is hazardous	III	Improve instructions
	Squeeze injury	Moving machine parts	II	Improve machine guards
Feeding new paper through machine	Operator falls, if paper tears	Heavy task, if brakes are not completely off, or if base roll is oval	II	Improve design for releasing brake pressure
	Squeeze injury	Paper and steel rolls rotating with great force	III	Develop automatic paper feeder, or change work routines (use the previous sheet of paper to pull through a new roll)
/Correction of disturbances	Squeeze injury	Disturbances occur often	III	Improve control system, and develop safer correction methods

Working with a paper rolling machine

5.4 Comments

Simplicity

The method is easy to learn. But, if the training period is too short, problems may arise as a result of the coverage of the analysis being inadequate. Similarly, the range of safety measures proposed may be narrow, compared with that generated when a more highly skilled person is involved.

Simple analyses can be conducted with little preparation and only a small amount of effort. But, if the work to be analyzed is more extensive or involves a lot of variation, then there is a need for the application to be more formal, and possibly for the assistance of experienced people.

An advantage of the method is that it is based on common job tasks and commonly accepted ideas on safety. For this reason, it is easy to teach the method and get it accepted for direct use by job supervisors and work teams.

That the method is based on a standard approach to safety matters may also be a disadvantage. It is harder to avoid having a blinkered view on the work involved.

Information materials

Information is needed to:

— Prepare the list of work phases.

— Identify the hazards.

Where systems have been in operation for some time, there is a body of experience available. Generally, this is possessed mainly by those who work directly with the equipment and by job supervisors. This knowledge can be accessed through a suitably-composed study team.

The information needed can also be obtained from:

1. Interviews.

2. Written job instructions (sometimes incorrect, always incomplete).

3. Machine manuals.

4. Work studies, if these are available.

5. Direct observation, the observer simply standing and watching.

6. Photographs, both to depict problems and to facilitate discussions within the study team.

7. Video recordings are valuable for tasks that are only seldom undertaken.

8. Accident and near-accident reports.

The list of phases of work

An important part of the analysis consists in producing the list of job tasks. Sometimes this can take longer than the identification of the hazards themselves. Only brief descriptions of the different phases are needed. It is more important that the list is sufficiently complete. This is checked before the identification stage is embarked upon.

One common dilemma arises when there is a major discrepancy between the job instructions and how the job is carried out. How is compatibility between the two to be achieved?

Time taken by the analysis

The time taken by an analysis may vary considerably, but the method can be regarded as relatively quick to apply. How much time is needed for any one analysis depends on:

— The magnitude/diversity of the task to be analyzed.

— The efficiency with which the analysis is conducted and the training of the participants.

A rule of thumb is that the identification of hazards takes 5 minutes per phase of work. The number of work phases may come to between 20 and 50. The identification stage of the analysis may therefore be estimated to take between one hour and half a day. The author's experience is that it takes roughly the same time to produce a list of work phases, and the same time again to conduct discussions on safety measures. In total, the analysis may take between half a day and two days.

6. Deviation analysis

6.1 On deviations

Systems do not always function as planned. There are disturbances to production, equipment failures, and people make mistakes. These are deviations from the planned and the normal. Deviations can lead to defective products, material damage or injuries to people. However, the majority of deviations are neither discovered nor corrected. They do not necessarily have a negative effect.

It is fairly easy to understand intuitively what deviations are and why they are important. In the analysis of hazards, a strict definition is not always necessary. Nor may it be desirable. The purpose of discussing deviations in an analytical context is to discover factors which might lead to hazards. The need for precise definition is greater in some cases, e.g. when a statistical classification is to be made. Deviations of one type or another appear in several methods for safety analysis. These are discussed in chapter 10.

There are a large number of terms in the area of safety which are used to denote deviations of one type or another. Some examples are disturbance, breakdown, fault, failure, human error and unsafe act. One example of a general definition is that a deviation is an event or condition which diverges from a norm for the correct and planned production process (Kjellén & Larsson, 1981). The norm is established by agreement among the people who occupy relevant positions within the company.

The deviation concept is a common element in a number of different theories and models. Kjellén (1984) has studied a variety of areas of application and definitions. On a general level, a system variable is classified as a deviation when its value lies outside a norm. Some system variables are:

— Event or act (i.e. part of a procedure or a human action).

— Condition (i.e. state of a component).

— Interaction between the system and its environment.

Examples of different types of norms that appear in the literature include:

— Legal; a standard, rule or regulation.

— Adequate or acceptable.

— Normal or usual.

— Planned or intended.

The relation between deviations on the one hand and increased risk and the occurrence of accidents on the other can be complex. An important component of some analyses, such as Fault tree or Event tree, is the examination and clarification of logical relations between deviations and accidents.

Many kinds of deviations can arise within a system. The consequences can be of a large number of different types, some leading to increased risk, others being harmless. Table 6.1 shows a classification of consequences of deviations. Consequences of types 1 and 2 are directly observable. The other types may be noticed directly in some cases, but in others they have to be looked for specifically. A further aspect of the classification concerns the extent to which the deviations can be corrected so that the system can be returned to a safe state.

Table 6.1 Classification of deviations by type of consequences

Consequence	Comments
1. Direct accident	The deviation leads directly to an accident. For example, the wing of an aeroplane falls off in mid air.
2. Potential accident	An accident occurs if people are in the danger zone or if the system is at a certain operational phase.
3. Indirect cause of accident	The deviation in itself does not lead directly to an accident but could start a chain reaction.
4. Contributing cause of accident	In addition to the deviation itself, certain other independent failures must occur for an accident to happen.
5. Increases the probability of an accident	The deviation in itself does not mean that an accident will occur, but it increases the probability of hazardous deviations.
6. Not dangerous	The deviation does not lead to an increase in risk.
7. Increases safety	The deviation increases safety. For example, someone does not follow an unsuitable job instruction, but works in a safer way instead.

6.2 Principles

Background

Deviation analysis is used to identify the deviations in a production system and the hazards to which these can give rise. From the beginning, the method was developed specifically for application to accident investigations. When applied in this way the analysis is called a deviation investigation. In this section, the discussion focuses mainly on common features of the two applications.

The method is partially based on a model of accident sequences (Occupational Accident Research Unit, 1979). This model involves the division of the course of an accident event into three phases:

1. During the initial phase a system starts to develop in an uncontrollable direction, as a result of the occurrence of one or several deviations.

2. During the final phase development is rapid, and a flow of energy is triggered off. The human being has few opportunities to influence events during this phase, but can attempt to escape or ward off the energy flow.

3. During the injury phase there is an injury to a human being.

Using this model, a near-accident is seen as a course of events which has a final phase, but does not have an injury phase. To be counted as a near-accident, the event should be perceived as hazardous. Thus, accidents and near-accidents can be described using the same model. The model was originally planned for use in the investigation of accidents that have actually occurred. It has been further developed for the analysis of the risk of accidents within a system, i.e. before an incident has occurred (Harms-Ringdahl, 1982, 1987b).

Basic tenets of Deviation analysis

A Deviation analysis is based on the following assumptions:

1. Production is a planned process, and the course of normal production can be defined.

2. Deviations can increase the risk of accidents.

3. A system consists of technical, human and organizational elements.

4. The risk of accidents is reduced if deviations are identified and can be eliminated or controlled.

We base the analysis on the supposition that production is a planned process and involves a normal sequence of events. A deviation is defined as an event or a state that diverges from the correct, planned or usual process. Deviations from good practice in planning and organization are also considered.

Check list for Deviation analysis

A check list, as shown in table 6.2, is used for analysis and investigations. It is designed as an aid for the identification of deviations, and is based on technical, human and organizational functions.

Comments on technical deviations

T1 General. These may be system disturbances basically caused by several possible alternative deviations. Deviations related to automatic functions and computer control are also included in this category. Examples include a failure to achieve a desired final result, that plant stops unexpectedly, or that a machine runs too quickly (see also table 6.3).

T2 Technical. These are deviations of a technical nature, e.g. the failure of a component or module, an interruption to energy supplies which may cause machinery to stop.

T3 Material. This applies to material that is used in the system, and also concerns transport and waste. Deviations can concern poor quality, wrong quantity, wrong delivery time, etc.

T4 Environment. This refers to abnormal or troublesome conditions of the indoor or outdoor environments. Examples include the accumulation of waste, faulty or dim lighting, bad weather, and other temporary environmental states which cause difficulties.

Table 6.2 *Check list for system functions and deviations*

Function	Deviation
Technical	
T1. General	Departure from the normal, intended or expected functioning of the system.
T2. Technical	Failure of component or module, interruption to energy supply, etc.
T3. Material	Poor quality, wrong quantity, wrong delivery date, etc.
T4. Environment	Waste, poor light, bad weather, any temporary disruptive state of the environment.
T5. Technical safety functions	E.g. machine guards, interlocks, monitors, which are missing, defective or inadequate.
Human	
H1. Operation/movement	Slip or misstep, error in execution.
H2. Manoeuvring	Lapse or mistake (error of judgement), e.g. choosing the wrong control button.
H3. Job procedure	Mistake, forgetting a step, doing subtasks in the wrong order.
H4. Personal task planning	Choosing an unsuitable solution, violations of rules and safety procedures and risk taking.
H5. Problem solving	Searching for a solution in a hazardous way.
H6. Communication	Communication error with people or the system, on either sending or receiving a message.
H7. General	Inadequate skills (physical or cognitive). Insufficient knowledge.
Organizational	
O1. Operational planning	Non-existent, incomplete or inappropriate.
O2. Personnel management	Inadequate staffing, lack of skills.
O3. Instruction and information	Inadequate or lacking. Absence of job instructions.
O4. Maintenance	Maintenance plans not followed.
O5. Control and correction	Inadequate.
O6. Competing operations	Different operations interfere with one other.
O7. Safety procedures	Missing, inadequate, disregarded.

T5 Technical safety functions. These are failures of devices designed to reduce risks, such as machine guards, interlocks and various types of technical monitoring equipment. Examples of deviations include devices that are defective or inadequate and equipment that has been removed or disconnected.

Comments on human deviations

As human errors are nearly always more complex by nature than technical failures (see sect. 2.2), it is more difficult to provide a simple classification of human deviations. This means that certain types of human deviations, e.g. forgetting something, or failing to take adequate safety precautions can be placed in all categories. For this reason, flexible use should be made of this part of the check list.

H1 Operation/movement. This applies to errors in the direct handling of material and equipment, e.g. slipping, falling over, missteps, etc.

H2 Manoeuvring. This category refers to indirect handling, using a machine or control system. Examples include choosing the wrong control button and manoeuvring an object in the wrong direction.

H3 Job procedure. This applies where there is a normal task procedure. Deviations comprise various types of mistakes (errors of judgement), such as forgetting a step, doing subtasks in the wrong order, misinterpreting signals, etc. Also, a person may totally abandon the normal job procedure and work by improvisation in stages.

H4 Personal task planning. This may allow several degrees of freedom and provide scope for several types of deviations, in the form of both mistakes and violations. An unsuitable solution may be chosen for a variety of reasons, such as inadequate knowledge or a lack of instructions. Violations, such as breaching regulations and risk taking, may have many different explanations (see sects 2.2 and 2.3). Not using personal protective equipment is a special case of this, but deserves special prominence.

H5 Problem solving. This is a complex activity, and possibly too advanced to be analyzed using this simple method. The reason for including this point in the check list is to make an attempt to identify situations where there is scope for a person to try to solve a problem in a hazardous way.

H6 Communication. This is an important component of many systems and job tasks. The category is used to identify situations where communications errors, either with another person or the system, can be hazardous. It covers misunderstanding/misinterpretation when information is received and the sending of erroneous or unclear messages.

H7 General. This category is conceived of as an extra check. It is used when considering whether the limitations of human beings may cause problems, either for themselves or the functioning of the system. It can apply where physical or cognitive skills are inadequate, or where knowledge is lacking on the job task or situation in question.

Comments on organizational deviations

The reservation mentioned above concerning human deviations also applies to this part of the check list. It should be applied flexibly.

O1 Operational planning. This is a general category which in principle also covers the points below. Planning can involve a variety of problems. Planning may simply be non-existent, or it may be incomplete or misguided.

O2 Personnel management. This is a matter of having the right person in the right place. Problems include a lack of staff, staff without the required skills, and a lack of plans for training or recruitment.

O3 Instruction and information. Those who do the job must have adequate information on how to do it in the right way and according to plan. This might apply to manuals for permanently-installed equipment or to job descriptions for occasional tasks. Problems include a lack of instructions, and instructions that are inadequate, out-of-date or simply incorrect.

O4 Maintenance. In many cases, details of relevance to safety will be important. Problems may include a lack of maintenance plans, plans that are not followed, important routine subsections that are missing, the unavailability of spare parts, and a way of working that is unsatisfactory.

O5 Control and correction. These are operations designed to ensure that equipment and activities function as planned. If they fail, the system should be returned to its "normal state" or plans should be modified as appropriate. Deficiencies in correction can lead to a steady reduction in the safety of both technical and organizational functions.

O6 Competing operations. This category refers to situations where different operations can have a disruptive effect on one another. The operations may be quite independent of each other, or they might compete for the same resources. For example, the number of cranes on a construction site is limited. If demand is great, a crane may not be available, causing workers to resort to hazardous manual lifting.

O7 Safety procedures. These are designed to ensure that hazards are identified and controlled in accordance with the norms that prevail at the workplace. The problem may be that activities are generally lacking or inadequate. Other deficiencies can include low or misguided priorities, unclear areas of responsibilities, poor routines and weak implementation.

Types of deviations

The check list above is structured in accordance with system functions, but deviations can be of many different types. Table 6.3 provides a summary of types of deviations which can be used for different functions. Some specialized methods, such as HAZOP and Action error analysis (chapts 7 and 10), employ similar check lists for types of deviations.

Table 6.3 *Types of deviations*

NONE
TOO LITTLE
TOO MUCH
WRONG TYPE
WRONG ORDER
WRONG PLACE
TOO LATE
TOO EARLY

Selection of deviations

The number of deviations in a system can be very large. Which are most important and how extensively they should be examined will depend on the aim and level of ambition of the analysis. It may be that the analysis is limited to deviations of types 1 to 5 in table 6.1. Alternatively, it might be restricted to one or some of the categories listed below. Deviations are divided into those that:

1. Can directly lead to injury.

2. Lead to the weakening or impairment of a safety function.

3. Require hazardous corrections in the course of production.

4. Seriously disturb production or make planning impossible.

5. Increase people's proneness to error (external disturbances).

6. Reduce control over the system.

Different levels of deviations

Some deviations directly lead to accidents, others lead to an increase in the probability that other deviations will arise, etc. Where there is a desire to work strictly with causal relations, the results from a Deviation analysis can be used as raw material for more precise forms of analysis, such as Fault tree analysis (chapt. 8).

Four types of safety measures

Deviation analysis includes a simple but systematic method for the generation of safety measures. To increase the safety of a system, improvements should be based on deviations that have been identified as hazardous. Efforts are made to generate measures which can:

1. Eliminate the possibility that a certain hazard will arise.

2. Reduce the probability that it will arise.

3. Reduce the consequences if it arises.

4. Result in the speedy identification of the deviation and provide for plans on how it should be corrected in a safe and effective manner.

6.3 Deviation analysis procedure

In a Deviation analysis, an attempt is made to identify in advance the deviations that may lead to accidents. The aim of the analysis is to obtain a relatively broad picture of the dangerous deviations within a system. The analysis usually includes a stage where proposals to increase safety are generated. Deviation analysis proceeds in a manner that is similar to Energy and Job safety analysis, and involves the same main stages.

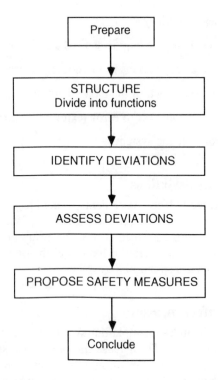

Figure 6.1 *Main stages of procedure in Deviation analysis*

Preparing

Preparations include defining how large a part of the system is to be covered by the analysis and specifying the operational conditions which are supposed to apply. At the same time, this determines what is not to be included in the analysis. A general piece of advice is not to be too restrictive in making this definition.

A second element in preparatory work is to ensure that the requisite information will be available during the course of the analysis (see sect. 6.6). As an aid to the analysis, a record sheet such as that shown in table 6.6 can be used.

Deviation analysis can be employed for problems other than that of accident risks. Before the analysis is embarked upon, a decision could be made on whether its scope should be widened to include other consequences, such as disturbances to production, poor product quality and damage to the environment (see sect. 13.5 on integrated approaches).

1. STRUCTURING

The aim of structuring is both to ensure that the entire system is covered and to divide it up into more elementary functions. The system is structured "functionally" on the basis of operations and activities. The result of structuring can be seen as a flow chart or as a model of the system.

The starting point is a description of operations. These are divided into blocks of an appropriate size. To cover general aspects, it may be a good idea to add a block denoted by such a heading as "General", "Planning" or "Organization". Principally, this acts as a reminder to include organizational aspects when the examination is conducted.

Here are some examples of how a structure can be established:

— On a production line, a number of production steps are taken in sequence. The different links in the production chain can be followed, and these can be divided up into different sections.

— For a transport system, a classification can be made in accordance with the various types of conveyors used.

— A series of actions needs to be taken (a procedure). One example from everyday life is that of preparing a meal. A structure is obtained simply by listing the various actions required.

Structuring can be seen as creating a model of the system on the basis of what happens. Usually, there are certain clearly evident main activities, and making the division does not present any problem. On the other hand, there may be a number of subsidiary activities which are not so immediately apparent. Moreover, some of these may be hazardous and must be included in the analysis. Examples of subsidiary activities include maintenance, the transportation of packaging material and the handling of waste.

When structuring is completed, a list of various sections or functions will have been obtained. These are then studied one at a time. Structuring is an important part of the analysis. It needs to be done with care, and a sufficient amount of time should be allowed for it.

2. IDENTIFYING DEVIATIONS

For each section, an attempt is made to identify deviations which can lead to accidents or have other negative consequences. A good starting-point may be to describe the purpose of the particular section under study. In searching for deviations, the check list shown in table 6.2 is used as an aid. For certain functions, e.g. for materials or procedures, the list of types of deviations shown in table 6.3 can be employed. The aim of this stage of the analysis is not to take up all conceivable deviations, as the total number can be very large. On the other hand, an effort is made to identify those which are critical for safety.

The analysis can then proceed with those deviations that are assessed to be particularly important. For example, if a certain component is critical, it is possible to continue with the analysis by looking in particular at Maintenance (O4) and Control and Correction (O5). Another example is where there are many possibilities for making an error in the handling of a machine and the skill of the operator is relevant to safety. Then, it can be important to look at Personnel Management (O2), Instruction and Information (O3), and perhaps also at Control and Correction (O4).

3. ASSESSING IDENTIFIED DEVIATIONS

The next step is to assess the seriousness of the identified deviations. The principles for this are discussed in chapter 11. In many cases, it is possible to obtain information through interviews or from the inspection of journals of operations. From accident investigations, data can be gathered on deviations that have involved a high level of risk.

4. PROPOSING SAFETY MEASURES

When the deviations have been assessed, an attempt is made to generate safety measures for those that are most important. At this stage, it is best to try to think as freely and creatively as possible, to develop a variety of ideas which can then be sifted and modified. All the four types of safety measures referred to at the end of section 6.2 should be systematically considered. The check list (table 6.2) of system functions, particularly its organizational section, can also be used as an aid. In concluding, the ideas are sifted through, and what emerges is put together into a proposal containing a number of different safety measures.

Concluding

The analysis is concluded by preparing a summary. This can contain accounts of the terms and conditions under which the analysis was conducted, the most important deviations and hazards, and the safety measures proposed.

6.4 Examples

One problem in giving practical examples is that a system needs to be described fairly extensively for it to be possible to understand how it operates and what can go awry. However, the two examples below provide rough outlines of how the analysis proceeds.

Example 1: A conference

This example is taken from an environment that is not particularly hazardous. On the other hand, a conference involves a "procedure" with which many are familiar. Most people will also have come across various examples of the disturbances and deviations that can occur.

Suppose that an important conference is to be arranged. The conference arranger is very concerned that everything will go well, as several previous conferences have gone badly. If something goes wrong, the arranger may be the object of ridicule. As a basis for planning, a Deviation analysis is conducted.

The structure (fig. 6.2) contains the most important activities and starts with the block "Planning", where other functions and the purpose of the conference can be taken up. The next step is "Invitation to conference", which involves attracting the interest of the correct target group, etc.

During the course of the analysis, the elements in the diagram can be divided up still further. Let us study the "Presentation" phase and divide it into its component parts. The first phase is "Speaking". The conference hall is large, so a public address system is needed. We assume that pictures or diagrams are to be shown. It is also important that the main aim of the presentation is included in the analysis. This is described as "Imparting knowledge".

Table 6.4 provides examples of what the analysis can generate for the subphases "Speaking" and "Showing pictures". There are a large number

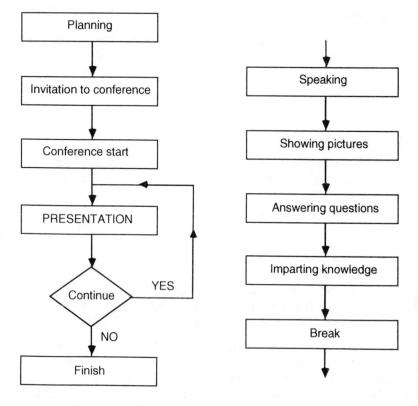

Figure 6.2 *Schematic description of a conference, and a division of the PRESENTATION phase into blocks*

of possible deviations even for these subphases alone. Obtaining an overall picture of possible disturbances and problems provides an opportunity for the better planning of the conference.

Example 2: Work with a computer-controlled lathe

This example concerns work with a computer-controlled lathe of a fairly conventional design. We assume that production runs are short, so that the product to be manufactured is changed from time to time. This means that lathe settings are adjusted, and tools and computer programs exchanged. There are a number of energies which can lead to

Table 6.4 *Deviations during conference presentation, coded in accordance with the types of deviations described in table 6.2*

Function	Deviation	Code
SPEAKING		
Public address system	Does not work (see below)	T1
	Oscillates, howls	T1
	Failure of a component	T2
Managing the public address system	Faulty adjustment	H3
	Presenter (P) cannot cope with the microphone	H7, O3
General	No checks on the sound equipment	O5
	Sound management not planned	O1
	No-one appointed to manage the sound	O2
	Inexperienced sound manager	O2
	Disturbance from drilling in an adjacent room	06
SHOWING PICTURES		
Projector	Wrongly adjusted projector, blurred picture	T1
	Faulty projection, picture too small	T1
	Several other possible technical disturbances	T1
	Defective lamp or fuse	T2
Pictures/diagrams	Text too small	T3
	Poor translucence	T3
	Wrong picture order	T3
	Upside-down/back to front	T3
Visibility	Strong main lighting, picture not visible	T4
	Weak main lighting, P cannot read script	T4
	Picture obscured by person or object	T1
General	P cannot switch on the projector	H7, O3
	New picture in wrong order	H2
	Failure to lower the blinds	O3
	Failure to turn on main lighting/spotlight	O3
	Failure to switch the lighting back on	O3
Planning	Lack of a projector	O1
	Lack of staff to give assistance when needed	O2
	Staff who lack the necessary skills	O2
	Inadequate instruction to staff	O3
	Projector not checked	O5
	No reserve materials	O4
	Equipment taken by presenter in an adjacent room	O6

serious injuries. One of the greatest hazards is that the lathe rotates at too high a speed.

Structuring

General structuring generates six main phases (fig. 6.3.) which jointly comprise the procedure for manufacturing any one of the products. At the "Setting-up" phase, the operator will tool the lathe, change settings, read in the computer control program, etc. At the "Testing settings" phase, the lathe is run for one program sequence, and then stops. The operator makes certain checks on the settings and adjusts the control parameters if needed. When the entire job cycle has been tested, automatic operations can then be set in motion. The testing phase itself contains several subphases which are also shown in figure 6.3.

Identifying deviations

Table 6.5 provides examples of different deviations that may occur in the course of "Setting-up" the lathe and "Testing settings". The examples come from a case study (Backström & Harms-Ringdahl, 1986), and all these deviations have occurred. Some can lead to accidents, while others either cause defects in the finished product or mean that extra time must be taken to complete the work. Table 6.6 shows how a Deviation analysis record sheet can be filled in.

There follows a few comments on some of the deviations. A "settings sheet" describes how the lathe should be set up, and which tools and control programs should be used. The deviation "Select wrong settings sheet" means that the machine will be set up for the wrong type of manufacturing. Which settings sheet is to be used is listed in coded form on the works order. An error can be made by the operator or might have occurred earlier in the chain.

The first part of the testing phase is for the operator to initiate the operational mode "TEST". The aim of selecting this mode is to run the job cycle sequence by sequence at low speed. The most fundamental deviation is that this mode is not selected. The lathe will then go directly into production mode, and the tool settings, etc. will not be properly checked. This gives rise to a major risk for accidents or breakdowns.

There are a number of deviations which may have this effect. On the

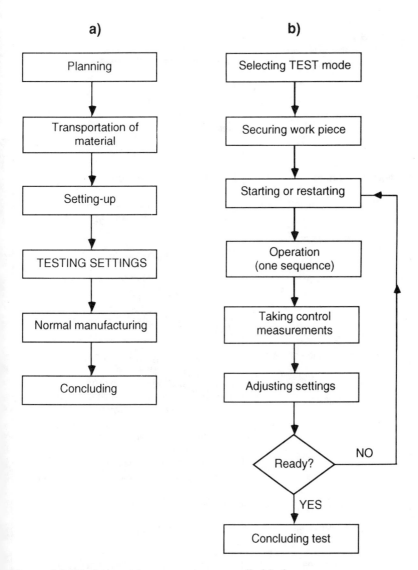

a)

Planning

Transportation of material

Setting-up

TESTING SETTINGS

Normal manufacturing

Concluding

b)

Selecting TEST mode

Securing work piece

Starting or restarting

Operation (one sequence)

Taking control measurements

Adjusting settings

Ready?

NO

YES

Concluding test

Figure 6.3 *Working with a computer-controlled lathe*
a) Production divided into blocks   b) The testing phase

Table 6.5 *Some deviations when working with a computer-controlled lathe*

Activity	Example of Deviation
SETTING-UP Studying works order and settings sheet	Select wrong settings sheet Wrong number on the settings sheet
Fitting cutting tools and accessories	Select wrong tool Fit tool in wrong place Fit tool incorrectly Defective or worn-out tool
Removal of tools	Leave the old tool in place
Reading in program	Error in loading procedure
Data transmission failure	Select wrong program (for another product) Select out-of-date version of program
TESTING SETTINGS General	Omit entire test procedure
Selecting TEST mode	Wrong indication of mode (lamp faulty) Press wrong button
Securing work piece	Inadequate fastening (technical or manual error)
Operation, one phase at a time	Excessive pressure of tool on work piece Work piece comes loose Speed of rotation too high (error in earlier installation, or technical failure) Work with safety hood open Operator puts head in machine to see better Unexpected machine movement (for the operator) Unwanted stop, e.g. with no READY signal Correction of disturbance in a hazardous way
Control measuring	Incorrect measurement
Adjusting settings	Wrong calculation Enter values incorrectly
Ready? (Continue test)	Finish the test before the entire cycle is completed (may depend on unclear information from the system)

Table 6.6 Extract from a Deviation analysis record sheet for working with a lathe, risks assessed as acceptable (0 or 1) and not acceptable (II or III) (see sect. 11.3)

Function / Part	Deviation	Hazard / Comments	Evalu-ation	Proposed measures
Testing settings	Omit procedure	Work piece can come loose at full operating speed	III	Instruct the operators on the hazards of omitting the procedure Extra interlock to avoid omission
/Selecting TEST mode	Wrong indication of mode Press wrong button	Work piece can loosen at full speed / If lamp is broken, a faulty indication is given	III	Change design of indicator
/Securing work piece	Inadequte fastening	Work piece can come loose / Many possible reasons; technical or manual failures	III	Conduct a Fault tree analysis to summarize possible failures and errors
/Operation, one phase at a time	Work with safety hood open Unwanted stop	Operator squeezed by tool or caught by rotating work piece Squeezed / If starts unexpectedly	II II	Restricted use, or interlock with e.g. dead man's grip when hood is open Develop safety correction routines Interlock with e.g. time out function for unwanted stops
/Control measuring	Incorrect measurement	Normally no danger, but the whole product batch can be destroyed	I–III	Choose an instrument which is easier to use Better illumination
/Adjusting settings	Wrong calculation Enter values incorrectly	See above	I–III	Better training of operators and good calculation facilities
/Continue test	Finish too early	Work piece can loosen / Possibly hazardous consequences at next stage	III	Clearer indication from the system when test procedure is completed

85

control panel, there is a button with the text "BLOCK DELETE". When the indicator light is on, the lathe is in automatic operating mode. If it is off, this means that "TEST" has been activated. Therefore, a defective lamp will mislead the operator. The control panel has a lot of buttons, labels and instructions, and there are many ways in which an operator can make a mistake.

The next phase involves securing the work piece. A large number of deviations may mean that the work piece is not fastened securely. Some of these may be related to the setting-up of the lathe during the previous phase. One conclusion may be that a special investigation needs to be conducted of the various deviations that will cause a work piece to come loose. Moreover, as there are so many, these should be summarized, e.g. using Fault tree analysis.

6.5 Accident investigation

Experience from Sweden is that accident investigations at companies can often be considerably improved. In the investigation of accidents there is usually a focus on the immediate course of events related to the injury. As a result, the safety measures that are proposed generally concern protective devices of one type or another. Frequently, no safety measures at all are generated. One explanation for this is that knowledge of the immediate course of events is not sufficient to establish which measures are needed. To get further, there is a need to study what happened before the accident, before everything started to happen quickly.

Deviation investigation is a method that can be employed. It is closely related to Deviation analysis, so the same check list can be employed. However, instead of searching for hypothetical deviations, an attempt is made to discover what preceded the accident in question. Near-accidents can also be investigated using the same method. Experience has shown that accident investigations conducted on the basis of deviations provide more information on the accident in question and generate a greater number of proposals for safety measures (Harms-Ringdahl, 1983; Kjellén, 1983).

In cases where the use of Deviation analysis is planned, there are some arguments for using related methods for both accident investigation and safety analysis:

1. Knowledge is obtained on which types of deviations have occurred at accidents in the company. This provides an aid for identifying relevant deviations and assessing their importance.

2. Experience of investigation improves the skills needed for analysis, and vice-versa.

The investigation of an accident should not be seen as a form of safety analysis. An investigation covers just a limited part of the system and only some hazards. Moreover, the "choice" of hazards investigated is determined randomly by the accident that has occurred. However, the investigation can be transformed into a safety analysis by using the accident itself as a point of departure, and then carrying out a deeper and more thorough investigation of the system as a whole. For this, either Deviation analysis or MORT (see sect. 9.2) can be employed.

Deviation investigation procedure

We imagine a situation where a preliminary investigation has been conducted, and this needs to be supplemented. The aim of the investigation is to generate safety proposals rather than to describe in detail what happened and the order in which it happened. The procedure is shown in figure 6.4.

1. SUMMARIZING THE COURSE OF THE ACCIDENT

This is achieved using information from the preliminary investigation. The starting-point is the accident event, which is then followed backwards in time (like rewinding a film in slow motion). Make a list of the deviations that arise.

2. IDENTIFYING DEVIATIONS

This stage involves supplementing the list of deviations. In principle, the analysis continues backwards until everything in the system is "normal". The relevant information can be obtained from the injured person, job supervisor, etc. In principle, two things are sought for: first, new deviations which have not previously been detected; second, follow-up information on those deviations which are already known.

87

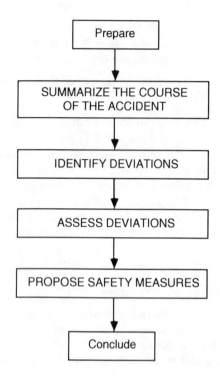

Figure 6.4 *Main stages of procedure in a deviation investigation*

As an aid for identification, the check list shown in table 6.2 can be used. It needs to be reformulated in an appropriate manner to be of help in conducting the interviews. For example, ask:

Was the machine working normally? (T1)
Had anything failed? (T2)
Was there anything unusual about the materials being used? (T3)
etc.

Questions on job supervision and planning may often be contentious, and are avoided in most investigations. But, organizational functions are important, and an investigation that does not consider them is incomplete. The check list may make it easier to pose questions in such a way that they are not perceived as being sensitive or loaded. Of course, the questions cannot have the exact form that they have on the check list, for example:

Were planning procedures followed? (O1, M3 and M4)
Was the planning adequate? (O1)
Were the job tasks of the injured person appropriate? (O2)

On whether any technical equipment had been defective:

Why was the fault not discovered? (O5)
Was the component included in the maintenance program? (O4)

3. ASSESSING THE DEVIATIONS

The next step is to assess the deviations. This can be done formally, as described in chapter 11, or simply by selecting the most important deviations for further treatment. These may be types of deviations which:

Occur frequently.
Are considered to be a problem.
Are in breach of regulations or company rules.

4. PROPOSING SAFETY MEASURES

Ideas for possible safety measures are based on the deviations selected. The measures are generated in the same way as when Deviation analysis is used (table 6.4). It is usually best to start with technical and individual deviations, and then investigate whether organizational conditions and routines can be improved. Try to think freely at the beginning. Probably, ideas for safety measures will come up even when attempts are being made to find explanations for why certain deviations have occurred.

The next step will be to assess the ideas for safety measures and give them more concrete form. In doing this, it may be that new and better ideas will be generated to replace some of the old.

Concluding

The investigation is concluded by making a summary recommendation for safety proposals. The summary might certainly contain proposals, appointing people to ensure that the measures are implemented and laying down a schedule for their implementation.

Some advice

We assume from the beginning that the purpose of the investigation is to generate proposals that will raise the level of safety, not to find scapegoats. It is important to explain this to the people interviewed. It both facilitates discussion and makes it easier to obtain information.

It is not always possible to find out exactly what happened. There may have been different alternative courses of events, or it can be uncertain whether a particular deviation really occurred. When searching for ideas for safety measures, these uncertainties need not be regarded as drawbacks. Rather, it is the opposite. If an accident can occur in several different ways, it is best that the safety measures cover all eventualities.

Example of an accident investigation

The accident occurred at a paper rolling machine in a paper mill. The example is taken from a case study discussed in section 15.8. The machine is used to cut up wide rolls of paper that come from a paper machine. The principle is that the paper from the large roll is wound up so as to pass through a system of rollers and reel cutters. The final result is a number of smaller rolls.

The accident

On the occasion of the accident, the injured person (P) was about to pull a new paper roll through the machine. His hand slipped and he received a severe cut from a reel cutter. The preliminary accident report stated that the paper had become crumpled, and when P pulled the sheet of paper down to correct the fault, his hand slipped. The supplementary investigation identified four further deviations (5 – 8).

Deviations

The summary recorded a total of eight deviations. The classification categories from table 6.1 are noted in brackets, but these will not usually appear in a summary.

1. P cut himself between thumb and forefinger, between the upper and lower reel cutters (injury event).

2. P's hand slipped (H1).

A new sheet of paper is drawn towards the reel cutters of a paper rolling machine

3. P tried to correct the faulty paper feed, but in a hazardous way (H2 or H4).

4. The paper became crumpled (T3).

5. The automatic equipment used to thread the paper through the machine functioned poorly (T1).

6. The work team departed from accepted practice. They should have started afresh, and threaded a new sheet of paper through the machine (O1, possibly H3 or H5, depending on the circumstances).

7. P was not aware of the safe job procedure when threading paper through the machine (H7).

8. P was an apprentice, working for the first day on normally-scheduled job tasks. It was his third day at the machine (O2).

Deviations 3, 6, 7 and 8 were selected for further investigation. The investigation involved:

1. Examination of the job introduction program for new employees.
2. Scrutiny of the job instructions for the cutting machine.
3. Checking whether the work team followed the instructions.
4. Listing the disturbances that occurred at the cutting machine.

It was found that the company had an ambitious job introduction program, but it was too much concerned with the company itself and too little with actual job tasks. The job instructions were relevant in principle but were phrased too generally. There were no instructions on what should be done when disturbances to operations occurred. The instructions available were followed but, as one person expressed it: "You couldn't really break the rules, anyway." The list of common operational disturbances was a long one.

Proposing safety measures

On the basis of this accident, the following safety measures were proposed:

1. The program for the introduction of new employees should be modified. Greater emphasis should be placed on occupational hazards and how to act when operational disturbances occurred. The introduction of a sponsorship system (under which each new employee would be supervised by an experienced worker) and the appointment of special instructors were also recommended.
2. The job instructions should be modified. A list of disturbances, showing what to do in each case, should be prepared.

These measures can be categorized as involving "Planning for the identification and correction of deviations". Similar disturbances to production were involved in other accidents. For this reason, a number of technical modifications were later implemented so as to reduce the likelihood that disturbances occurred.

6.6 Comments

General

Deviation analysis is a relatively new method in safety work. One reason for including it in this book is that it has received a favourable reception from those who have learnt and employed the method. It provides a new perspective for safety engineers and offers an opportunity for them to tackle problems they have seen but not known what to do about.

Deviation analysis is more difficult than Energy or Job safety analysis. However, it is possible to select a degree of detail adapted to the skills of the analyst and to the type of system which is to be analyzed. Usually the level selected is such that the analysis can be conducted in a few hours or during one day.

The method is general by nature and is not only applicable to accidents. Many undesired events are preceded by deviations. For example, the principles have been applied to interruptions to production, accidents leading to environmental harm, and fire and explosion hazards.

Planning and information

Most production systems are more complex than they seem at first sight. Information on the system and the problems that arise in use can be obtained from:

— Direct observation.

— Written descriptions and drawings.

— Interviews.

— Accident reports, operations records, etc.

A practical way of conducting an analysis is to form a study team. The team should contain people who are acquainted with technical functions, how the work is organized, and how the work is carried out in practice. Then, there will be adequate access to information on the system and its problems. The number of conceivable deviations in a system can be considerable. For this reason, a practical way of proceeding is to make a preliminary assessment of the deviations which might conceivably occur.

Structuring

It is primarily the structuring of the object under study that is regarded as difficult. Using Energy or Job safety analysis (or any technically-oriented method) a structure can generally be found more-or-less immediately. Some of the difficulties are that:

1. There is seldom a ready-made structure available at a suitable level of detail. It is the job of the analyst to divide the system into functions.

2. Descriptions of system functions are often incomplete.

3. There are often several different ways of dividing up the system.

In addition, it can be hard to assess in advance the degree of detail required. On some occasions, general functions are enough; on others, details are needed. In the two examples given in section 6.4, a general classification is made first, and then some of the classified functions are broken down in greater detail.

This means that the task of describing and structuring a system can take longer than the identification part of the analysis. Careful structuring can be of benefit in other ways, providing, for example, the basis for the improved planning of previously-neglected parts of the system, or for the production of job instructions.

That there are several possible solutions to the problem of structuring need not be a drawback. Several different solutions can be used, thus making the analysis more comprehensive. Some blocks may be identical, but they need only be analyzed once.

In principle, the situation is simpler when a system is at the planning stage. Then, it is possible to work on the basis of the plans alone. In the case of systems that are already in existence, there are discrepancies between what is planned and what takes place in practice. Moreover, one is confronted with a more complex reality. By its very nature, planning is incomplete, and systems are modified as they develop.

Identifying deviations

The check list of deviations may seem a long one. But, the purpose of the list is to identify deviations. It is not meant to represent a "template" or model that should be rigidly applied. In practice, there is no time to ponder over each item at great length. It is shown by experience that users of Deviation analysis tend not to make extensive use of the check list particularly after a certain period of time. It becomes natural for them to observe and search for deviations without it.

Information on deviations that have occurred in relation to previous accidents is valuable. A number of accident reports from the system covering a period of several years can be gathered together. Supplementary investigations can be conducted in conjunction with a person who is acquainted with these cases. This provides a list of deviations which have actually led to accidents, and which can be used at the identification stage. It can also increase the motivation of the study team, if it can be shown concretely that deviations do lead to accidents.

One specific comment should be made in relation to human deviations. The analysis is not principally designed to examine human behaviour. It is concerned with the possibility of human error and the consequences that errors might have. If there are problems, solutions can be sought in improved technical solutions, organizational planning, training, etc.

Different levels of deviations

As mentioned previously, there are many types of deviations, and they can have different consequences. Some lead directly to a serious event, but depend in turn on other deviations having previously occurred. Others affect the likelihood of the appearance of further problems.

During the identification stage, a deviation can be recognized as one of the following:

— A consequence of a deviation that has previously occurred.

— One that leads to a deviation that is already known.

In such cases, reference is simply made to the deviations already identified. At the end of the analysis, the material can be structured so that related deviations are merged. Another way is to provide a summary, e.g. in the form of a fault tree.

In the course of the analysis, a list of different deviations is prepared. This involves generating a one-dimensional description of a two-dimensional phenomenon. This may seem cryptic, but will be clearer to those familiar with Fault tree analysis, which can be regarded as being a two-dimensional description.

7 Hazard and operability studies

7.1 Principles

In the chemicals processing industry, there is often a potential for major accidents. There is also a tradition that hazards are identified and control measures taken. A number of authorities and organizations work with these issues. One method that has become well established in the chemicals industry is HAZOP, an abbreviation for Hazard and operability studies. Extensive guidelines have been prepared on how the method should be employed (CISHC, 1977; Lees, 1980; ILO, 1988).

The basic idea behind HAZOP is that a systematic search is made for deviations that may have harmful consequences. The HAZOP technique is designed to stimulate the imagination of designers in a systematic manner, thus enabling them to identify conceivable hazards.

The model of the system employed is a technical process model. Hazards are defined as deviations which could cause damage, injury or other forms of loss.

HAZOP's several characteristic elements are defined as follows.

INTENTION — A specification of "intention" is made for each part of the installation to be analyzed. The intention defines how that part of the installation is expected to work.

DEVIATION — A search is made for deviations from the intended which might lead to hazardous situations.

GUIDE WORD — Guide words on a check list are employed to uncover different types of deviations.

TEAM — The analysis is conducted by a team, comprising members with a number of different specializations.

This first section provides an account of guide words, while the second describes the stages of procedure used for HAZOP. In section 7.3 a simple example is provided. The chapter concludes with some comments and tips, principally obtained from the original HAZOP specifications.

Table 7.1 *Guide words in HAZOP*

Guide word	Meaning
NO or NOT	No part of the intention is achieved. Nothing else happens.
MORE	Quantitative increase, e.g. in flow rate or temperature.
LESS	Quantitative decrease.
AS WELL AS	Qualitative increase. The intention is fully achieved, plus some additional activity takes place, e.g. the transfer of additional material (in a conveyance system).
PART OF	Qualitative decrease. Only a part of the intention is achieved.
REVERSE	Logical opposite of intention, e.g. reverse direction of flow.
OTHER THAN	Complete substitution. No part of the original intention is achieved. Something quite different happens.

Guide words

One of the most characteristic features of **HAZOP** is the use made of "guide words". These are simple words or phrases which are applied to the "intention" of either a part of an installation or a process step. Guide words can be applied to:

— Materials.

— Unit operations.

— Layouts.

A simple example

The simple example that follows illustrates the use of guide words. The example refers to a liquid which is to be pumped into a pipe.

The first three guide words are immediately and easily understandable. NO means that nothing is pumped, **MORE** that more liquid than intended is pumped, LESS that less than intended is pumped; AS WELL AS means that something in addition to the intended pumping of the liquid takes place. AS WELL AS might refer to:

— The liquid containing some other component, e.g. from another pipe.

— The liquid also finding its way to a place other than that intended.

- A further activity taking place at the same time, e.g. the liquid starting to boil inside the pump.

The guide word PART OF means that the intention is only partially realized. If the part of the installation under study is designed to fulfil more than one objective, perhaps only one of these is met:

- A component of the liquid is missing.
- If the liquid is to be supplied to several places, only one of these receives its supply.

The guide word REVERSE denotes that the result is the opposite of what is intended. In the case of the liquid, this might be that the flow is in the reverse direction.

The guide word OTHER THAN means that no part of the original intention is realized. Instead, something quite different occurs. The guide word may also mean ELSEWHERE. In terms of the example, OTHER THAN might be due to:

- The pumping of a liquid other than the liquid intended.
- The liquid ending up somewhere other than intended.
- A change in the intended activity, e.g. that the liquid solidifies (or starts to boil) so that it cannot be pumped.

7.2 HAZOP procedure

The stages of procedure in HAZOP are extensively described in the literature referred to above. A rather simplified description is provided here. When using HAZOP, all stages are usually applied to each part of the process, taking the parts one at a time. An example of the special record sheet used for HAZOP analysis is shown in table 7.2.

Preparing

The aim of the analysis has to be specified. It may be to examine the proposed design of an installation or to increase the safety of an existing plant by generating improved job instructions. The types of hazards to be considered can also be specified. These can concern hazards faced by people at the installation, product quality, or the influence of the plant on the surrounding environment.

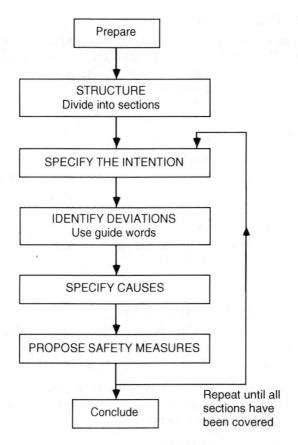

Figure 7.1 *Main stages of procedure in HAZOP*

A boundary for the analysis is set by specifying which parts of the installation and which processes are to be analyzed. A team is appointed to conduct the analysis. As usual, preparation also involves the gathering of information and planning for the implementation of the study.

1. STRUCTURING

The installation is divided up into different sections. In the case of a continuous process, the division is into tanks, connecting pipes, etc. The analysis is then applied separately to each section, one at a time.

2. SPECIFYING THE INTENTION

The intention of each part to be analyzed is defined. This specifies how it is thought that the part will function. If the designer participates, he or she can provide an explanation. Otherwise, it will be the person who is most familiar with the installation.

3. IDENTIFYING DEVIATIONS

Using the guide words, an effort is made to find deviations from the specified intention. The guide words are applied one at a time.

4. SPECIFYING CAUSES

For each significant deviation, an attempt is made to find conceivable causes or reasons for its occurrence.

5. PROPOSING SAFETY MEASURES

The consequences of the deviations are examined. The possible seriousness of these should also be assessed. Matters of the assessment and grading of consequences are not taken up in the HAZOP manuals.

For deviations that may have serious consequences, an effort is made to find control measures. The person responsible for the implementation of each measure can be specified on the record sheet.

The safety measures may apply to:

— Changing the process (raw materials, mixture, preparation, etc.).

— Changing process parameters (temperature, pressure, etc.).

— Changing the design of the physical environment (premises, etc.).

— Changing routines.

6. REPEATING THE PROCEDURE

When analysis of a section of the installation is completed, this is marked on the drawing. The next part is then analyzed, and the procedure continues until the entire installation has been covered.

Concluding

The analysis is concluded by preparing a summary, but further follow-up might be needed. This might include liaising with those responsible for the control measures, further development of the safety proposals, etc.

7.3 Example

The HAZOP analysis illustrated in figure 7.2 is a simplified and amended version of an example originally presented by the UK Chemical Industries Association (CISHC, 1977). It concerns a plant where the substances A and B react with each other to form a new substance C. If there is more B than A, there may be an explosion.

We begin with the pipe, including the pump, that conveys material A to the tank. The first step is to formulate the INTENTION for this part of the equipment. Its aim is to convey a specific amount of A to the reaction tank. In addition, the pumping of A is to be completed before B is pumped over.

We apply the first guide word, NO or NOT. The deviation is that no A is conveyed. Possible causes of this are sought for, and it is easy to come up with several conceivable explanations:

1. The tank containing A is empty.

2. One of the pipe's two valves (V1 or V2) is closed.

3. The pump is blocked, e.g. with frozen liquid.

4. The pump does not work, for one of a variety of possible reasons. The motor might not be switched on, there might be no power supply to the motor, the pump might have failed.

5. The pipe is broken.

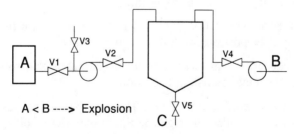

Figure 7.2 Schematic description of an installation

The consequence of the deviation is serious, and involves the risk of explosion.

The next guide word is MORE. The deviation means that too much A is conveyed. Reasons for this might be that:

1. The pump has too high a capacity.

2. The opening of the control valve is too large.

The consequences will not be as serious as above. C can be contaminated by too much A, and the tank can be overfilled.

The third guide word is LESS, meaning that too little A is conveyed. Reasons for this might be that:

1. One of the valves is partially closed.

2. The pipe is partially blocked.

3. The pump is generating a low flow, or is operating for a shorter time than intended.

The consequence may be serious.

The fourth guide word is AS WELL AS. The deviation is that A is conveyed, but that something else happens. Examples of such deviations are that:

1. A further component is pumped through the pipe, which might be due to:

   — Valve V3 being open, resulting in another liquid or gas entering the flow.

   — Contaminants in the tank.

2. A is pumped to another place as well as to the tank. This might result from a leak in the connecting pipe.

3. Another activity is taking place which competes with the pumping. Would it be possible for A to boil in the pump?

The consequence of all these different deviations is that too little A is conveyed, meaning the risk of explosion.

The fifth guide word is PART OF. The deviation is that just a part of the intention is fulfilled. It might be that a component of A is missing, although this appears not to be possible in this case.

Table 7.2 Record sheet for a HAZOP analysis

Guide word	Deviation	Possible causes	Consequences	Proposed measures
NO, NOT	No A	Tank containing A is empty V1 or V2 closed Pump does not work The pipe is broken	Not enough A, explosion	Indicator for low level Monitoring of flow
MORE	Too much A	Pump has too high a capacity Opening of V1 or V2 is too large	C contaminated by A Tank overfilled	Indicator for high level Monitoring of flow
LESS	Not enough A	V1, V2 or pipe are partially blocked. Pump gives low flow, or runs for too short a time	Not enough A, explosion	See above
AS WELL AS	Other substance	V3 open, air is sucked in	Not enough A, explosion	Flow monitoring based on weight
PART OF	—			
REVERSE	Liquid pumped backwards	Wrong connection to motor	Not enough A, explosion, A is contaminated	Flow monitoring
OTHER THAN	A boils in pump	Temperature too high	Not enough A, explosion	Temperature (and flow) monitoring

The sixth guide word is REVERSE. This would mean that liquid is conveyed from the reaction tank to the container for material A. Conceivable deviations include:

1. The pump is operating in reverse. This would occur if the power supply was wrongly connected to the motor.

2. Liquid is running backwards from the reaction tank or the connecting pipe due to gravity.

The consequence can be serious.

The seventh guide word OTHER THAN means that no part of the original intention is fulfilled. Instead, something quite different occurs. Some examples of such deviations are:

1. A liquid other than the intended liquid is pumped.

2. The liquid finds it way to some other place.

3. There is a change in the intended activity. It might be that the liquid solidifies or starts to boil, so that it cannot be pumped.

This is a highly simplified example. It is not specified whether a continuous or batch process is involved, how the quantities of A and B are controlled, etc. Table 7.2 shows the analysis summarized on a HAZOP record sheet.

7.4 Comments

Hazard and operability studies are well established and a large amount of collective experience has been accumulated. The reference literature (CISHC, 1977; Lees, 1980; Kletz, 1983; ILO, 1988) offers plenty of advice on how analyses should be used, planned and conducted. Many of these pieces of advice reflect the more general viewpoints on safety expressed in chapter 12 of this book. Given, however, that they are based on experience of just this method, a fairly extensive account is also given in this section.

The basis for a safe installation lies in the application of an accepted, well-established technique for which a long period of experience is available. Different specialists can contribute their own specialized knowledge. In addition, there are different norms and directives from the authorities that must be applied.

All installations have their own special features, and hazards can manifest themselves in various ways. This is why there is a need to conduct an investigation of each installation to find specific faults and items that have been neglected in its design.

It should have emerged from this account that HAZOP does not provide an automatic means for the discovery of hazards and the generation of safety measures. The results obtained, as is the case with most analytical methods, are dependent on creative thinking. *Good results come from the application of a systematic approach, utilization of the guide words and the formation of a team with a suitable membership.*

When is HAZOP used?

HAZOP can be used in different situations:

— At the planning stage, before detailed design and construction decisions are made.

— Before system start.

— For an existing installation.

At the planning stage

The greatest benefit is obtained if an analysis is conducted in conjunction with the design of the installation. The optimal point in time is when decisions are being made on how the plant is to be constructed and detailed design documentation is ready (design freeze). Drawings that are sufficiently detailed for the analysis are then available.

A HAZOP analysis takes time. In the case of a large installation, it can be a matter of several months, even if several teams work in parallel. It is possible either to embark on the construction work and accept the risk of changes or to wait until the analyses are ready. But construction schedules must allow time for this.

Before system start

There can be a point in conducting the analysis even when the installation has been nearly completed and when instructions for users have already been prepared. The reasons why an analysis is justified at this stage are:

- Important changes have been made.
- Operating instructions are critical to safety.
- The new installation is similar to one that already exists. The changes primarily affect the process and not the equipment.

For an existing installation

An installation where safety was adequate at the time when operations were started may deteriorate over the years. A series of changes may have meant that different types of hazards have arisen. This particularly applies if safety issues were not carefully considered when the changes were made. It may also be that sufficient attention was not paid to safety at the design stage, or that requirements for operational safety have become stricter over time.

Information

The literature on the method stresses the importance of the availability of a sufficiently detailed documentary base for the analysis to be conducted. This means, among other things, that a HAZOP analysis cannot be conducted at too early a stage of the planning of an installation. A start on the study can only be made when detailed documentation is available.

Drawings and instructions must be up-to-date and correct. Drawings often need to be updated, which can require a substantial amount of effort. In the case of existing installations, it is often found that information is incorrect.

The study team

Guides to how HAZOP should be conducted stress the importance of working as a team. This applies to team composition, skills and attitude. HAZOP does not replace knowledge and experience. If the team lacks either of these, the results will be unusable. It is important that members of the team have a positive and constructive attitude towards their task. Success depends on the ability of the participants to think constructively and with imagination. Members must be selected with care, and motivation must be promoted.

The team should not be too large, containing a maximum of seven members. A study requires members with different specialized forms of technical expertise, with knowledge covering the process, measuring and control techniques, etc. The team must have sufficient knowledge on how the installation is designed to function. Thus, even for reasons of efficiency alone, it is important to be able to obtain answers directly, and avoid having to guess or to obtain information from outside sources.

If the team contains members who have the authority to make direct decisions on changes, this makes the study more effective. If the installation is being designed or constructed by an outside supplier, representatives of both user and supplier should participate in the analysis. The work requires continuity, so members should only be replaced in cases where absolutely necessary.

The role of the team leader is important. He or she must be familiar with the HAZOP method, capable of leading the discussions, and able to ensure that the schedule for the analysis is followed. The task of the leader also involves producing the documentation needed for the analysis. It is sufficient if it is the leader alone who has thorough knowledge of the method. Other members can participate without extensive training. A training period of between one hour and two days has been mentioned, depending on the level of ambition of the study. The leader must ensure that proceedings at meetings run efficiently and agendas are kept to. There must not be so many delays that the members get bored with the analysis. The leader summarizes the results when each section of the study has been completed. He/she also marks the drawing after, for example, a pipeline is ready.

Time taken by the analysis

For most installations, a HAZOP analysis is time-consuming. For this reason, proper scheduling is required. The average period of time required by an analysis is 10 – 15 minutes, either per component or per activity covered by a job instruction. This means one to three hours for each main unit, e.g. a reactor with several connecting pipelines. For the analyses to be effective, study meetings lasting three hours at most are recommended. Moreover, these meetings should not take place more than twice or three times a week.

Thus, if the object to be analyzed is a large one, careful planning is required. Planning involves the following:

- Finding time for the entire object.
- Getting through the meetings in a reasonable amount of time.
- Having the necessary information material available at the meetings.
- Ensuring that time is available for control measures and follow-up activities decided upon at the meetings.

To complete the analysis in a reasonable time, several teams working in parallel may be required. One of the team leaders should then adopt the role of coordinator.

Safety measures

The finding of safety solutions can be conceived of in terms of two extremes. In practice, there will be a compromise between the two:

- A solution is produced after each source of risk (hazard) is discovered.
- No solutions are produced until after all the guide words have been applied.

The team is often predominantly, or even exclusively, composed of technicians. In such cases, it must be remembered that not all problems need to be solved by making technical changes.

The follow-up of measures and other such activities are important. Who is responsible is noted on the record sheet. If the analysis is conducted at the planning stage, or in the case of a new installation, there should be a readiness to make changes. In the case of an existing system, measures are needed so that the system will function better than before.

Analysis of batch production

The literature already referred to contains supplementary advice for the study of installations where batch production takes place. In addition to drawings of the plant, information is needed on the sequence in which the procedure is carried out. It may be either automatically or manually controlled. The information material may consist of job descriptions, flow sheets, etc. A summary description of the settings of valves, etc. may be needed for different situations that can arise

The analysis can be structured so as to follow job procedures rather than different parts of an installation. The same guide words as before are employed, although these can be re-formulated as appropriate. For example, EARLIER and LATER may be employed for time or job sequences. When applied in this way, the method is similar in certain respects to Deviation analysis.

Miscellaneous

Taylor (1979) has suggested a variant of HAZOP in which the emphasis is on physical variables. The analysis is then based on a check list that covers temperature, pressure, etc., and a simplified set of guide words is applied to these.

Some companies wish to receive detailed documentation of the analyses. This is sensible in itself, but involves a significant amount of extra work. According to Kletz (1983), it is seldom that such information is made use of afterwards.

If a lot of changes are made after a HAZOP study, a new round of analyses may be required. The additional study would be designed to discover whether new problems had been introduced by the changes already implemented.

Experience has shown that problems of start-up, close-down, etc. are often neglected by over-specialized design groups working in isolation. Sometimes, a guide word MISCELLANEOUS is employed to capture deviations or problems that have not been identified using the other guide words. The category is primarily designed to cover occasional activities which can lead to problems. Examples include starting up and closing down the plant, inspection, testing, repairs, cleaning, etc. This guide word does not have a natural place within HAZOP, but can be valuable for the detection of further problems.

8 Fault tree analysis

8.1 Introduction

A fault tree is a graphical representation of logical combinations of causes that may lead to a defined undesired event or state. Examples of types of final events are an injury to a human being, failure of equipment, the release of poisonous gas and an interruption to production.

Fault tree analysis is perhaps the best known method employed in safety analysis. It started to be used in the 1960s. The method is perhaps of greatest value for complicated technical systems where a functional failure can have serious consequences and also where considerable resources can be allocated for hazard analysis. The method is relatively difficult and is generally used by specialists. There is an extensive literature on the method (e.g. Lees, 1980; Henley & Kumamoto, 1981; IEC, 1990), and a number of computer programs are available to aid the design of fault trees and make calculations.

It can be questioned whether the method is appropriate for common safety work outside high-risk sectors. But a general knowledge of Fault tree analysis is useful even for those who will not use the method directly. The purpose of the description given here is to acquaint the reader with the method and provide a basis on which simpler kinds of fault trees can be constructed. However, probabilistic estimates, which form an important area of application in Fault tree analysis, are only briefly referred to. To some extent, the traditional focus of Fault tree analysis has been extended. As this short account goes beyond purely technical factors, human actions and the taking of control measures are also considered.

Some of the advantages of Fault tree analysis are:

1. It is an aid for identifying risks in complex systems.

2. It makes it possible to focus on one fault at a time without losing an overall perspective.

3. It provides an overview on how faults can lead to serious consequences.

4. For those with a certain familiarity with the analysis, it is possible to understand the results relatively quickly.

5. It provides an opportunity to make probabilistic estimates.

Some of its disadvantages are:

1. It is a relatively detailed and, in general, time-consuming method.

2. It requires expertise and training to conduct.

3. It can provide an illusion of high accuracy. Its results appear advanced and, when probabilistic analyses are conducted, these can be presented in the form of a single value. As with most methods, there are many possible sources of error.

4. It cannot be applied mechanically and does not guarantee that all faults are detected. In general, different analysts will produce a variety of different trees. A tree can have different forms and still have the same content.

5. Its implementation generally requires detailed documentary material to be available.

8.2 Principles and symbols

A basic concept is that of event, which can also denote the state of an installation or an extreme value of a variable. In Fault tree analysis an either/or approach is adopted. Either an event occurs or it does not. The event statement can then be designated as "true" or "false". This can also be expressed in terms of the logical values "1" and "0", meaning that binary logic and Boolean algebra can be applied.

This is both a strength and a weakness. The approach has the advantage that faults in complex systems can be described in a simple manner. The weakness is that many differences of degree that exist in reality cannot be taken account of by the analysis.

In designing a fault tree, a set of symbols is used. The set has a number of variants and only a limited selection of symbols is taken up here. Symbols in Fault tree analysis are of two kinds: gates and events. The most important are shown in figure 8.1.

The first three symbols refer to "events" that describe a fault of some kind. They might, therefore, be better described as "failure events", or in other such terms. They may be events in a strict sense, i.e. something that happens, but may also refer to a faulty state, e.g. a component that has failed.

The conditional symbol is used to show how normal conditions or

Symbol	Designation	Function
 ◯	BASIC EVENT	Basic event or failure
 ◇	UNDEVELOPED EVENT	Causes are not developed
▭	EVENT	Event resulting from more basic events
 ⬭	CONDITIONAL EVENT	Event that can occur normally
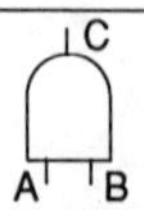 AND gate symbol with C on top, A and B inputs	AND gate	Output event occurs only if all input events occur simultaneously
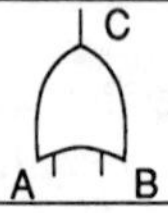 OR gate symbol with C on top, A and B inputs	OR gate	Output event occurs if any one of the input events occurs
 △	TRANSFER SYMBOL	Represents an event which comes from another lower-order fault tree or which is to be transferred to a higher-order tree

Figure 8.1 Symbols used in Fault tree analysis

events can also affect the system. Sometimes, such a symbol is used in combination with a special gate called **INHIBIT**. The transfer symbol is used to divide a tree into several smaller parts.

The AND and OR gates are used to provide logical connections between the various events. A somewhat more extensive description is provided in section 8.4.

Example of a fault tree

The appearance of a fault tree may be illustrated by a simple example.

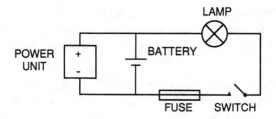

Figure 8.2 Example of a lamp circuit

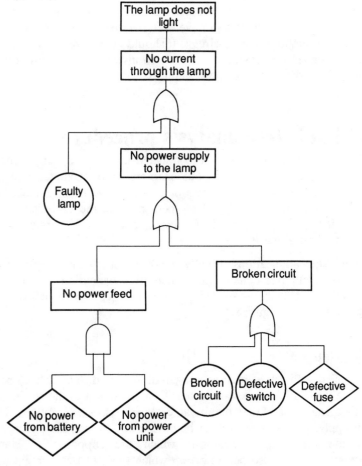

Figure 8.3 Fault tree for a lamp circuit

113

A lamp is connected into a circuit, as shown in figure 8.2. The lamp is fed by a power unit and there is a battery to provide reserve power in case the unit fails. A fault tree is wanted to analyze the case where the lamp does not light when switched on.

The top event is that the lamp does not light. This is because there is no current through the lamp. In turn, this may be due to the lamp being faulty or there being no power supply to the lamp. The power feed will fail if both the power unit and the battery fail to operate.

The tree contains three basic events, and there are also three "undeveloped events". That the fuse is defective may be due to aging or some other factor. But it might also have been overloaded as a result, for example, of a temporary short-circuit. It should be possible to develop this further. Similarly, it should be possible to investigate why power is not coming from the battery or power unit.

8.3 Fault tree analysis procedure

A Fault tree analysis cannot be conducted in such a direct manner as the analyses described in previous chapters. Constructing a tree is as much an art as a straight-forward building operation. Success very much depends on the ability of the person doing the analysis. For this reason, it is difficult to provide a universal and clear description of how one should set about it.

Probabilistic applications usually involve four main steps: system definition, fault tree construction, qualitative evaluation and quantitative evaluation. One variant, with emphasis placed on the construction stage, is shown in figure 8.4.

Preparing

As is usual with safety analysis, pre-conditions need to be defined before the analysis itself can be conducted. Constructing a fault tree involves detailed analysis and may require an extensive set of assumptions. These may apply, for example, to the boundaries of the system under study and the operational conditions which are supposed to prevail. Assumptions may also be needed on which types of faults might occur and which should be excluded.

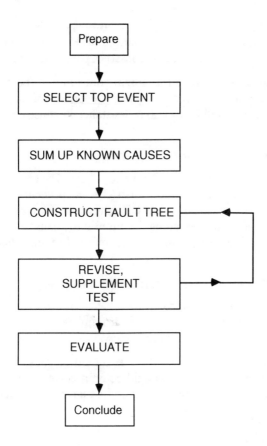

Figure 8.4 *Main stages of procedure in Fault tree analysis*

1. SELECTING THE TOP EVENT

The first step is to select the undesired event to be analyzed. This should be carefully defined. If a top event is too broadly defined, it can probably be divided up into several different events. A separate fault tree can then be constructed for each case.

2. SUMMING UP KNOWN CAUSES

When constructing the fault tree, existing knowledge of faulty states and failure events should be utilized. It facilitates the analysis if a prelimi-

115

nary examination of the failures that may arise is conducted. Alternatively, the results of a Deviation or HAZOP analysis can be used. This material can be used to construct part of the tree.

After this step, a list of faults that might contribute to the occurrence of the top event will have been obtained. Generally the list will not be complete, but it will still be of major assistance in constructing the tree.

3. CONSTRUCTING THE FAULT TREE

Construction of the tree begins with the top event. The first step is to consider whether it can occur in more than one independent way. If so, the system has to be divided up using OR gates. The analysis continues by progressing downwards, searching for more basic causes. Some of these can be obtained from the preliminary list referred to immediately above.

4. REVISING, SUPPLEMENTING AND TESTING

Construction is a trial-and-error process. Progress towards a better and more complete tree is made in stages. A number of rules of thumb for carrying out this work are provided below.

It is hard to know exactly when the tree should be considered complete. No important causes of failure should be omitted. A first check is to see whether all the points on the preliminary list have been covered.

5. EVALUATING

The completed tree is then evaluated and utilized. Depending on the purpose of the analysis, a number of different steps can be included at this stage. Some of these are discussed more extensively in section 8.4.

— *Direct evaluation of the result.* The tree provides a compressed picture of the different ways in which the top event might occur. It also provides a picture of the barriers (safety functions) that exist. The tree shows if some failures can directly lead to the occurrence of the top event.

— *Preparation of a list of minimum cut sets.* As shown in section 8.4, a cut set is a collection of basic events which can give rise to the top event. A minimum cut set is one which does not contain a further cut set within itself.

— *Ranking of minimum cut sets.* Combinations of failures to which special attention should be paid can be evaluated and ranked on the basis of the minimum cut sets.

— *Estimation of probabilities.* If information on probabilities for the bottom events is available, or if these can be estimated, the probability of the occurrence of the top event can be calculated from the list of minimum cut sets.

Rules of thumb

In constructing a fault tree, the rules of thumb shown in table 8.1 can be utilized. Rules 1 – 7 are applied in the course of constructing the tree. Rules 8 – 10 are used from time to time to test whether the tree has a valid logical structure. The list is partly based on the account provided by Henley and Kumamoto (1981). A further source is the author's experience of problems encountered by beginners when they first embark on Fault tree analysis.

Figure 8.5 provides examples of how some of these rules might be used. Rules 6 and 7 have been merged into one. There is one OR gate with

Table 8.1 Rules of thumb for constructing and testing a fault tree

1. Work with concrete events and states. These should be denoted in the form of event statements that are either true or false.
2. Develop an event into a further event that is more concrete and basic.
3. Divide an event into more elementary events (OR gate).
4. Identify causes which need to interact for the event under study to occur (AND gate).
5. Link the triggering event to the absence of a safety function (AND gate).
6. Create subgroups frequently, preferably dividing into pairs.
7. Place a heading above every gate.
8. Do not make tacit assumptions, and do not let preconceived opinions (non-explicit assumptions) control the analysis.
9. Think logically and in terms of structure. Do not mix up cause and effect.
10. Test the logic of the tree from time to time during construction. Start from events at the bottom of the tree and suppose that these occur. What will the consequences be?

four inputs. This has been divided up into several gates, but the tree still has an identical logical function. If any of the basic events occur, then event statement A is true. Such a subdivision makes for more systematic analysis. The disadvantage is that the diagram takes up more space.

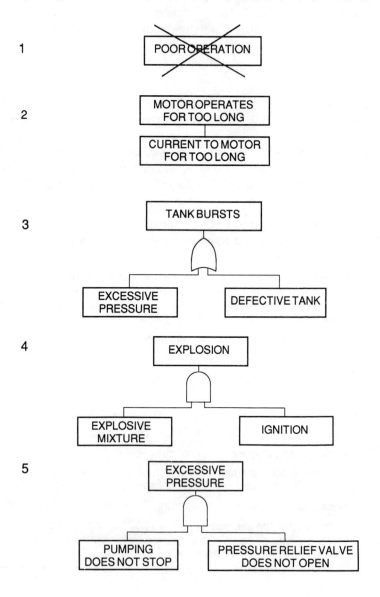

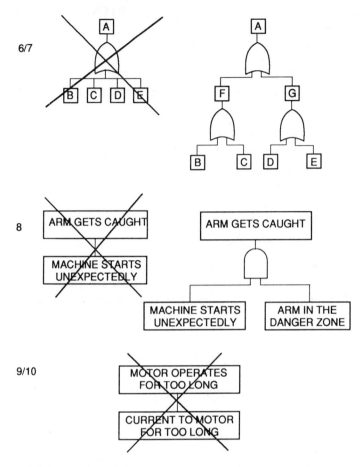

Figure 8.5 *Examples of the application of rules of thumb in Fault tree analysis*

The example of Rule 8 shows a line of thought that has been neglected. That the machine starts unexpectedly will only lead to an accident if a person is directly in the danger zone.

The example for merged rules 9 and 10 shows a case of the confusion of cause and effect. Suppose that the motor operates for too long. This does not lead to current flowing for a long time. In this case, the mistake is obvious, but in more complicated contexts it is easy to perform such logical somersaults.

8.4 More on Fault tree analysis

General

This section takes up a number of additional themes related to Fault tree analysis, such as types of symbols, other kinds of trees and forms of evaluation.

Traditional applications of Fault tree analysis involve the making of quantitative estimates. In such cases the main stages are as follows:

1. Definition of the system.

2. Construction of the fault tree.

3. Qualitative evaluation.

4. Quantitative estimation.

Defining the system is often the most difficult part of the analysis (Lees, 1980). It is important to be well acquainted with the system and how it functions. Its physical boundaries need to be specified and further conditions may need to be defined. This can apply to:

1 The top event.

2. Initial conditions.

3. Failures which might be supposed to occur.

4. Excluded events.

Only a few accounts of how a fault tree is constructed in practice are available. Perhaps the most extensive is that of Henley and Kumamoto (1981), several of whose ideas have been utilized in preparing the list of rules of thumb above. The authors themselves point to the lack of practical guides and suggest that the construction of a fault tree is as much an art as a science. Two analysts will not construct identical trees. (But this also applies to safety analysis in general as soon as one goes beyond a trivial level.)

For certain special types of systems, more extensive advice on the construction of trees can be provided. For example, a formal method for the synthesis of fault trees for electrical systems has been developed (Fussel, 1973). The procedure is based on the examination of the system's individual components. Mini-trees are constructed on the basis of individual component failures. These are then put together to form a composite fault tree.

Large numbers of computer programs of different types are available as aids for the construction of fault trees. For a beginner, it seems best to start by constructing a tree by hand. Otherwise, there is a serious risk that the task of managing the program will be predominant and analytical thought neglected.

On the use of logical symbols

The most important symbols are shown in figure 8.1 above, but they can be presented in alternative ways. The functions can also be expressed in the form of logical expressions or as truth tables. Figure 8.6 shows how these different forms of presentation are related.

Let us start with the logical variables A and B and their corresponding event statements, which can be "true" or false". If A is "false", it is given the value 0; if A is "true", it obtains the value 1. Examples of what A might represent are: *motor switched on* or *safety device removed.*

AND and OR gates may have an arbitrary number of inputs. For AND gates all input statements must be true for the output statement to be true. For OR gates it is enough for just one of several input statements to be true for the output statement to be true.

A further function has been added, i.e. the negation (NOT). This is not used in Fault tree analysis, but still requires some attention. Negation statements take the form: if A is true, then Z is false.

A truth table shows how a logical function depends on the input variables. This can be explained through the examples given in figure 8.6.

— NOT (Z) is a function of a variable (A). When $A = 0$, $Z = 1$. For $A = 1$, $Z = 0$.

— The AND gate (X) has two inputs. For $A = 1$ and $B = 1$, $X = 1$. For other combinations of A and B, $X = 0$.

Probability equations have also been included in figure 8.6. Fault tree analysis is frequently employed to provide a basis for probabilistic calculations, and it is appropriate to introduce some of the formulas here. The probability that A shall occur within a certain time interval is

121

FUNCTION	AND	OR	NOT
Symbol	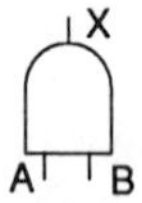 (AND gate symbol, output X, inputs A, B)	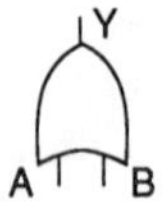 (OR gate symbol, output Y, inputs A, B)	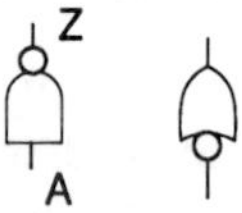 (NOT symbols, output Z, input A)
Alternative symbol	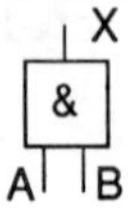 (box labeled & , output X, inputs A, B)	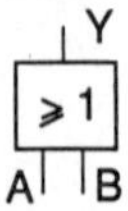 (box labeled ≥1 , output Y, inputs A, B)	
Function denotation	$X = AB$ $(X = A \cdot B)$ $(X = A \cap B)$	$Y = A + B$ $(Y = A \cup B)$	$Z = A'$ $(Z = A^*)$ $(Z = \bar{A})$
Truth table	A: 0 1 B 0 | 0 0 B 1 | 0 1	A: 0 1 B 0 | 0 1 B 1 | 1 1	A Z 0 | 1 1 | 0
Probability	$p(X) = p(AB) = p(A)\,p(B)$	$p(Y) = p(A{+}B) = p(A) + p(B) - p(A)\,p(B)$	$p(Z) = 1 - p(A)$

Figure 8.6 *Different ways of describing logical relationships*

denoted as p(A). The probability that A shall not occur is $1 - p(A)$. For the formulas for p(X) and p(Y) given in figure 8.6 to be applicable, it is assumed that A and B occur independently of each other.

Things that break

An installation that breaks does so as a result of its load being greater than its strength. For a more extensive description, see O'Connor (1981). Normally, installations are designed and constructed so that there is a margin between lowest strength and highest load. If a failure occurs, this may be because the margin is too narrow. Let us take a bridge as an example.

This is illustrated in figure 8.7. The load is not constant but varies with time. Sometimes there are a lot of vehicles on the bridge, at other times

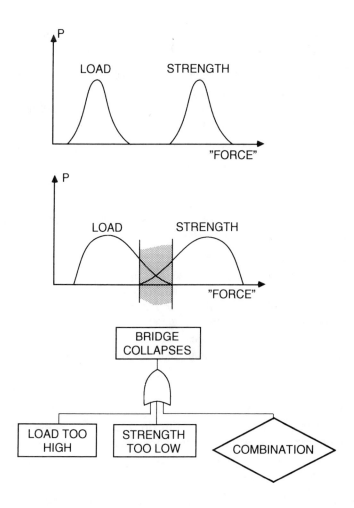

Figure 8.7 *Relationship between probability and strength. The shadowed area shows that the construction may collapse*

there are a few or none. The curve on the left shows the probability (p) of the bridge being exposed to a certain load.

Nor is the strength constant. The bridge can rust, or severe cold may mean that it is weaker at certain times. If a large number of identical bridges have been built, it is not certain that all have equal strength. Construction errors or material defects can arise. For this reason, a curve is needed, which shows the probability that the bridge will cope with a certain force.

In the case of the upper of the two diagrams, there is a margin between load and strength. How large the safety margin should be is decided at the design stage. For example, the relation between maximum permissible load and strength might be set at a factor of ten.

The lower diagram shows a situation where the margin is insufficient. Sooner or later the bridge will collapse.

A fault tree can be marked to denote that load is high relative to a certain specified value, or that strength is lower than this value. A combination of these faults can also arise. Sometimes it can be difficult to distinguish between the two cases.

Component failures are often classified as primary failures, secondary failures and command faults (linked using OR gates). Primary failures occur during normal operating conditions, e.g. from the effects of natural aging. Secondary failures occur when a component is exposed to conditions for which it is not designed. Command faults refer to functions where the component does work but where its function cannot be fulfilled, e.g. as a result of signals that are faulty or absent.

Other types of trees

Relationships between different functions can be described by other types of trees, not just fault trees. Three such trees are shown in figure 8.8. It is easy for those unaccustomed to Fault tree analysis to mix up different types of trees.

An organization can be described as a tree. Figure 8.8 shows an example of a "hierarchical" tree, showing the order of relations between departments. Such trees can also be used to describe technical systems.

A classification into subgroups or classes can also be illustrated by a tree. The word taxonomy is used to describe the classes created when there is a strict classification. Such a tree is not a fault tree, but can form a part of one. It can be used to distinguish between different events which may have the same final result.

A success tree can be used to describe what is required for an installation to work. Such trees are also described as logic flow or function diagrams. They are the opposite of fault trees, which show what is required for something not to function.

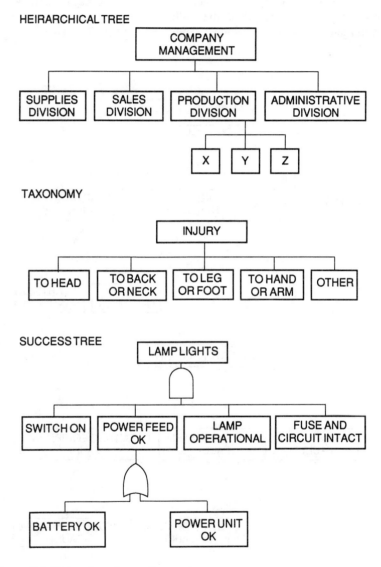

Figure 8.8 *Examples of tree diagrams*

Relationship between fault and success trees

There is a close relationship between fault and success trees. The lighting of a lamp is described as a fault tree in figure 8.3 and as a success tree in figure 8.8. In the fault tree, the functions are negative: they concern what is defective. For the bulb to light, everything must work (AND gates). For the bulb not to light, it is enough for there to be just one fault (OR gate).

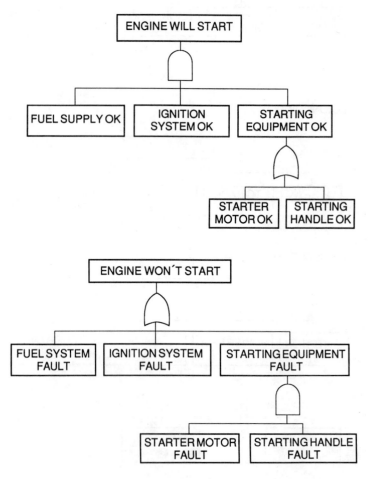

Figure 8.9 *Example of transforming a success tree into a fault tree*

There is a general way of transforming a success tree into a fault tree. De Morgan's theorem lays down that:

(AB)' = A' + B', and

(A + B)' = A'B'

The theorem can be proved by setting up truth tables for the right and left sides of the equations. It can then be observed that the tables are identical in all positions. The theorem can be generalized to cover more than two variables.

This can be expressed in words as a simple rule. A success tree can be transformed into a fault tree by:

1. Negating all statements, i.e. writing the opposite.

2. Transforming all AND gates into OR gates.

3. Transforming all OR gates into AND gates.

It is a mistake, however, to consider that a fault tree can be constructed simply by transforming a success tree in accordance with this rule. A fault tree must go more deeply into the "negative" side, taking account of different types of faults in depth. Moreover, it is seldom that the top event will be of the type "System does not work". Nevertheless, the way of thinking described above may be of assistance. It can be applied to either a complete system or a subsystem. An example of a transformation is provided in figure 8.9.

Simple evaluation

A fault tree can be used as a basis for making probabilistic estimates, but it is also possible to draw direct conclusions from studying the tree. Some of the questions raised in such an evaluation are as follows:

— Are there basic events which directly lead to the top event?

— In preparing the analysis, have assumptions been made which exclude failures that may have a striking effect on the level of risk? These assumptions can be clearly specified or may be implicit. For example: Has it been supposed that electrical power will be supplied the whole time?

— Can common cause failures occur? This would mean that faults which are supposed to be independent are triggered by the same event (see sect. 2.1). Examples include the failure of the power supply or that several human errors arise in sequence as a result of poor instructions or the misinterpretation of a situation.

— Have the system's safety features been clearly shown? They appear as AND gates.

— Can the level of safety be increased? AND gates, which represent different kinds of safety functions, can be inserted in the tree where they provide the greatest benefits. An example is the introduction of control routines for certain operations or for when certain system states arise.

Dividing the tree into minimum cut sets

As a basis for further evaluation, the tree is often divided up into minimum cut sets. A cut set is a collection of basic events which can give rise to the top event. A minimum cut set is one which does not contain a further cut set within itself. Computer programs are available that can provide assistance. In the case of simpler trees, a division into cut sets can be carried out by hand (Fussel, 1976; reproduced in Lees, 1981).

Ranking of minimum cut sets

A qualitative comparison can be made between cut sets, with the aim of revealing which basic events make the greatest contribution to the occurrence of the top event (e.g. Brown & Ball, 1980). A ranking of this sort can be based on two separate factors. The first is the number of basic events that are included. If there is just one, the set has greater significance than where two or more are included. The second factor concerns types of faults, which can be ranked as follows:

1. Human error (HE)

2. Active component failure (AC)

3. Passive component failure (PC)

The rank order of the different minimum cut sets would then be:

HE; AC; PC; HE & HE; HE & AC; HE & PC; AC & AC; AC & PC; PC & PC etc.

With simpler trees, this ranking can be carried out directly without there being a need to produce minimum cut sets.

Quantitative estimates

A fault tree can be used for estimating the probability of the occurrence of the top event. Usually this is based on the minimum cut sets. Estimates of probabilities for all the bottom events of the tree are needed. An approximation of the probability of the occurrence of the top event is derived by summing the probabilities of the minimum cut sets. This presupposes that these probabilities are low.

An alternative estimation procedure involves working directly from the tree, moving upwards stage-by-stage and applying formulas for the AND and OR gates. This provides a clearer picture of which types of faults make the greatest contribution to the occurrence of the top event. The correctness of the result depends on two conditions: that bottom event failures are independent of one another, and that each bottom event appears in only one place within the tree.

The reader wishing to go further into estimation methods in Fault tree analysis should refer to the more specialized literature. There are various computer programs available which will help with the calculations. The greatest general problem, however, is to find failure data of sufficient quality on the various components of the system. Despite the difficulties, probabilistic estimates can provide major benefits. For example, they enable solutions to be compared and provide assistance in setting control priorities. Lees (1980) summarizes some of the problems involved in using Fault tree analysis as a tool for estimation:

— The fault tree may be incomplete. There is no guarantee that all faults and all logical relationships will be included.

— Data on probabilities are lacking or incomplete.

— Estimates for systems with low failure probabilities are difficult to verify.

8.5 Example

System description

Figure 8.10 shows a sketch of a chemicals processing plant. In the container, two chemicals react with each other over a period of 10 hours and at a temperature of 125°C. When the reaction is complete, the contents are tapped off into drums through the opening of a valve. If the temperature exceeds 175°C toxic gas is formed.

The two chemical ingredients are pumped over from two other tanks. The volumes pumped are read off on two special instruments. The contents of the tank are heated by a heating coil controlled by a relay. The temperature rises at a rate of approximately 2°C per minute when the heating device is connected, and falls at roughly the same rate when it is off.

The temperature is measured using a sensor. The signal from the sensor is linked to the relay and forms a part of the temperature control circuit. If the temperature is lower than required, the relay switches the heating on. If the temperature is too high, the heating is turned off by the relay.

The signal from the sensor is also connected to an alarm which is activated if the temperature exceeds 150°C. If the alarm sounds, the operator is supposed to switch off the power feed.

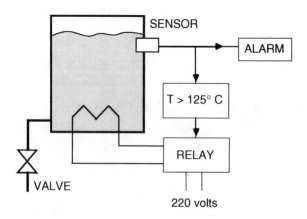

Figure 8.10 *Heating chamber with alarm facility*

Preparing

A fault tree for the event that poisonous gas is formed is wanted. It is based on the system description above. The level of accuracy of the description is low, but it is sufficient for a preliminary analysis. We assume that a proposal as specified in the sketch has been made, and that the analysis is designed to evaluate this proposal.

Selecting the top event

The proposed top event POISONOUS GAS IS FORMED can be used directly.

Summing-up known causes

No previous investigation has been held. It can be seen directly that there are some possible faults, which may be hazardous. Let us look at some examples:

— Sensor out of order, giving a low temperature reading.

— Circuit controlling the heating device out of order (does not cut out).

— Alarm circuit failure.

"Sensor out of order" will lead to a hazardous situation in this context. Other failures can arise, but these mean that the temperature will be too low. Such failure events are not included in the fault tree.

Constructing the fault tree

We start with the top event and see that it is caused by the heating element operating for too long a time. The analysis might then continue in several different ways. We decide to divide the system into two parts, i.e. before and after measurement of the temperature. Two failure events are then relevant:

1. The information on temperature is incorrect at the output of the monitoring circuit.

2. The heater cut-out does not work (despite receipt of a correct signal).

We now have the upper part of the fault tree (fig. 8.11), and go on to

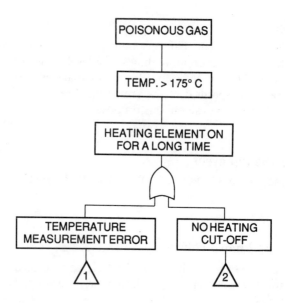

Figure 8.11 *The upper part of the fault tree*

consider temperature measurement error (fig. 8.12). This might be due to a fault in the measuring circuit, but no details of the circuit are available. So, we draw in a rhombus to mark an undeveloped event. That the measured temperature at the sensor is too low may have a large number of different causes. The most obvious is that the level of the liquid is below the sensor.

The part of the tree that shows how failures to cut out the heating device may arise is shown in figure 8.13. We do not have much information on its technical design, so the fault tree is rather small. That the alarm does not interrupt the system may have several causes. We do not know how the alarm will be installed or which people should take action when it sounds. Despite this, some types of general failures can be marked in.

Two protective actions have been assumed. One is that the operator has been instructed to make a report if deposits appear on the sensor. The other is that the operator should stop the operation if the liquid does not reach the sensor. These protective actions have been considered in the fault tree. But, as the tree describes failure events, they are

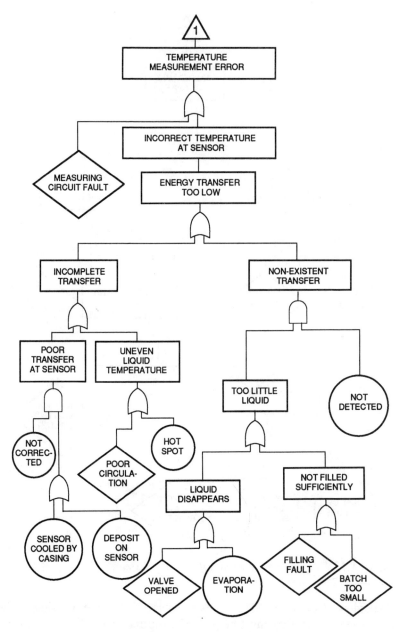

Figure 8.12 Fault tree for temperature measurement error

133

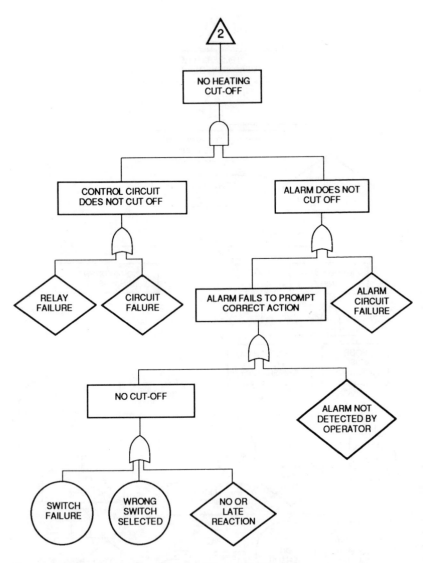

Figure 8.13 *Fault tree for cutting off heating device*

marked as "Not corrected" and "Not detected". For the functions to be of any real value, they must be part of a routine check. For this reason, they can be considered to be rather weak.

A general comment on the design of the tree is that there is too little space on the tree itself for adequate descriptions of failure events. The significance of each event will have to be made clear by its context and by surrounding events. For this reason, an explanatory list of the events on the tree may be needed.

Revising, supplementing and testing

The tree has been developed though a process of revision and supplementation. It would require too much space to show how this takes place. One simple test is to check for the inclusion of the "known causes" produced at the beginning of the analysis.

Evaluating

The tree includes three failures that lead directly to the formation of poisonous gas. One is a measuring circuit fault. The other two are related to the temperature being uneven within the chamber. This might be supposed to be due to "Poor circulation" or a "Hot spot" at the heating element.

Let us prepare a simple ranking list of minimum cut sets directly from the tree. We adopt the principles described in the previous section using the factors: Human error (HE), Active component failure (AC), and Passive component failure (PC).

The tree contains three purely technical failures which on their own can lead to the occurrence of the top event. There are also three combinations of two human errors which lead to an accident. Moreover, there are a number of other combinations of two faults which lead to the top event.

There remains one further source of failure, which has not been marked on the tree. This concerns the situation where the container has not been emptied completely. If the heating is switched on by mistake, the temperature will be too high. Such a situation might to some extent be fitted into "Liquid disappears", but it is not fully covered. The analysis only treated the system during continuous operation.

Afterwards, it is possible to state that the range of the analysis was too narrow. The greatest risks may lie in the heating being on when the procedure is started or stopped.

Table 8.2 *Combinations of events ranked according to their importance*

Combination of events	Comments
AC 1. Measuring circuit fault 2. Poor circulation 3. Hot spot	Single failure of active component
HE & HE 4. Valve opened & Not detected 5. Filling fault & Not detected 6. Batch too small & Not detected.	Double human error The operator releases some liquid, and does not realize that it may be hazardous. Too small an amount of liquid has been inserted from the beginning. Too small a batch of the chemicals has been ordered by job supervisors.
HE & AC 7. Deposit on the sensor & Not corrected 8. Sensor cooled by casing & Not corrected 9. Evaporation & Not detected 10–15. Control circuit does not cut out & Alarm does not lead to cut-off	Human error and failure of active component There is a deposit on the sensor, meaning that the recorded temperature value is too low. Moreover, the operator does not remove the deposit. Possibly a failure that arises on repair, as the sensor is not adequately insulated from its casing. Moreover, it is not detected and corrected. The quantity of liquid diminishes, e.g. through evaporation. Also, this is not detected. Six combinations are possible. Not listed here.
AC & PC 16–19. Failures in control and alarm circuits.	Failure of active and passive component Four combinations of technical failures in the control and alarm circuits are possible. Not listed here.

The conclusions drawn from the tree are as follows:

1. Three single failure events may lead directly to the occurrence of the top event. The safety level has to be regarded as unacceptable.

2. The installation requires radical re-design and an increased level of safety.

3. The new design and construction proposal will require analysis. Both what precedes and follows the heating phase must be considered.

The tree differs in several respects from a conventional tree. Above all, this applies to the bottom events. In many cases, they are not failures in that they refer to something that is out-of-order or has ceased to function. Rather, they can be characterized as departures from what has been designed or planned. Thus, they are equivalent to "deviations" in the sense employed in Deviation analysis or HAZOP. To go further and attempt to estimate probabilities, the tree must be more stringently constructed. Some of the bottom events must be defined more precisely and the relations between them further specified.

There are ten rhombuses at the bottom of the tree. Some of these refer to technical failures which could be developed further if more technical information was available. Other rhombuses are used to indicate that an event is worthy of further study

Eight items at the bottom have been ringed, marking faults that are fundamental. However, it is not obvious which bottom symbols should be chosen with this simple type of tree. Several of the ringed sections could be further analyzed in new trees.

8.6 Comments

Learning Fault tree analysis

Fault tree analysis is the most difficult of the methods presented so far. Its disadvantages and advantages have been described in section 8.1. After proper training it is not difficult to conduct an analysis. But it requires effort and may take a long time. If hazards with major consequences are to be studied, using Fault tree analysis is justified.

The length of time it takes to learn the method obviously depends on level of ambition and previous knowledge. It may be fairly easy for electronic engineers and computer programmers who are trained in the handling of logical circuits and functions.

137

Some problems

One problem is that a tree may be large and require a large amount of time. A second problem is that the analysis may have too great a focus on technical failures. Human and organizational factors may be neglected. For this reason, immediate concentration on technical failures alone should be avoided. If these prove to be extremely important, they can be studied more deeply at a later stage of the analysis.

A simplified picture

A fault tree is a simplification of reality in many senses. One concerns the adoption of the either/or approach, another the setting-up of strict logical connections between events. This means that some information must be excluded. Perhaps these problems are greatest if there is a desire to go beyond a conventional technical application.

Using existing trees

It may be that a fault tree is already available, and the analyst has the task of evaluating and interpreting it. A fault tree can be seductive, encouraging the belief that it is complete and has been well thought through. What might have been overlooked? Some questions that can be posed are as follows:

— Which assumptions about and simplifications to the system have been made?

— Are only technical failures included?

— Does the tree accord with the "rules of thumb"? (Those given in table 8.1 are in no way generally accepted or applied. But they do provide some sort of measure of quality.)

— Are there only OR gates? (If all failures will lead to the occurrence of the top event, there are grounds for wondering whether the tree is correct.)

A variety of applications

The fault tree methodology can be applied for a variety of purposes, which are not mutually exclusive. They include:

- Making probabilistic estimates.

- Identifying alternative chains of events that might lead to an accident.

- Compiling and providing a logical summary of results from other analyses, such as Deviation analysis and HAZOP. It can be used to make a transition between one-dimensional and two-dimensional descriptions.

- Organizing information from the investigation of an accident that has occurred. It can provide a means for examining the results obtained.

In many cases it can be fruitful to construct a tree with an accident that has occurred as the top event. Some of the reasons for this are as follows:

- Motivation for making use of the results is stronger. When a serious accident has occurred, the top event is no longer an abstract one. Instead, it is something of direct relevance.

- The compact description offered by a fault tree places emphasis on the overall picture and not on particular details.

- In some cases, there is uncertainty over which deviations and failures actually occurred. Instead of viewing uncertainty as a weakness, it can be used to show alternative ways in which accidents may occur.

- Broadening the range of the investigation and searching for deeper explanations become part of a natural process.

9. Some further methods

9.1 Introduction

The previous chapters have contained descriptions of five methods for safety analysis. These are methods that can be employed to analyze accident risks in the course of normal safety work. But they represent only a small number of the conceivable methods available. To broaden the picture, this chapter provides short accounts of some further methods. To some extent, these have been selected arbitrarily. Nevertheless, they have been chosen so that a wide range of different ideas are represented.

The methods taken up are:

MORT (Management oversight and risk tree).

THERP (Technique for human error rate prediction).

Action error analysis.

FMEA (Failure mode and effects analysis).

Event tree analysis.

Cause-consequence diagrams

Change analysis.

MORT is a method that focuses on the safety work of organizations. Action error analysis and THERP are two examples of analyses of human error, the former qualitative, the latter quantitative. FMEA and Event tree analysis are two well-established methods for safety analysis. Change analysis is an example of how a slightly different angle on hazards can be obtained.

The chapter also includes a section on methods described as quick (or rough) analyses. These are methods, or ways of working, which are systematic to a greater or lesser extent, but do not meet the requirements of thoroughness demanded from a proper safety analysis.

What is not included in this summary are the different hazard indices that have been developed for a variety of purposes. Probably the best known of these is the Dow Fire and Explosion Index which is used in the chemicals industry (Dow Chemical Company, 1976).

9.2 Management oversight and risk tree

Organizational activities govern how an installation is designed, how work is carried out, who works at the plant, what safety routines there are, etc. The quality and focus of these activities have decisive importance for the existence of hazards and how risks are controlled.

For this reason, it is important to have methods available for the analysis and assessment of the safety work of organizations. At the same time, it is a difficult subject — for a variety of reasons. Organizations and activities are not tangible objects, and it is not easy to get a grip on them. Written documentation reveals only a part of the reality. What causes difficulty is that there are informal decision making paths, people with varying views on what is relevant, etc.

These themes are also discussed in the chapter on Safety Management (chapt. 13), but there is a special method for the analysis of organizations which fits naturally into this section. This is called Management oversight and risk tree, abbreviated to MORT.

The origin of Management oversight and risk tree analysis lay in the need for an in-depth method for accident investigations. Development of the method dates from 1970. Its area of application has since been extended. A detailed guide and an account of the reasons for using MORT have been prepared by Johnson (1980). There is also a rather more summary description available (Know & Eicher, 1976).

"MORT emphasizes that when an accident reveals errors, it is the system which fails. People operating a system cannot do the things expected of them because directives and criteria are less than adequate. Error is defined as any significant deviation from a previously established or expected standard of human performance that results in unwanted delay, difficulty, problem, trouble, incident, accident, malfunction or failure" (Johnson, 1980). The energy model is an important element in MORT.

The MORT tree

The MORT logic diagram can be seen as a model of an ideal safety program. It can be used for:

—   The investigation of an accident.

—   The analysis of an organizational program for safety.

141

The MORT tree is a general problem description. It is rather like a fault tree and the same symbols are used. The tree contains around 200 basic problems. But, if it is applied in different areas, the number of potential causes it describes can rise to 1 500. For this reason, only a sketch of the method is provided here.

The top event, for example, may be an accident that has occurred. This can be due to an "assumed" hazard or to an "oversight or omission" (the two main branches of the tree), or both.

For a hazard to be "assumed", it must have been analyzed and treated as such by company management. Thus, the combination where a certain type of accident tends to occur and no specific control measure has been taken is not sufficient for the hazard to be counted as assumed.

The other main branch of the tree takes up organizational factors and is called "Oversights and omissions". It has two subsidiary branches, one of which is called "Specific control factors" and which focuses on what occurred during the accident. This is further divided into the accident itself and how its consequences are reduced, e.g. through fire fighting, provision of medical treatment, etc. The second subsidiary branch treats "Management system factors" and focuses on the question "Why". It is divided into three further elements: policy, implementation, and risk assessment systems.

The various elements in the tree are numbered. These numbers refer to a list which is provided as a complement to the tree. For each element there are specific questions which the analyst should pose.

Assessment

The analysis involves going through the elements in the tree and making an assessment of each. There are two assessment levels: "Satisfactory" and "Less Than Adequate" (LTA). Assessments are in part subjective, i.e. different people may make different judgements. Nevertheless, the availability of a list of specific, and often concrete, questions for each element reduces the degree of subjectivity.

Procedure

The analysis is conducted by following the MORT chart, first in general and then in greater detail. Questions for which it is possible to find direct answers are marked. Colours are used to code the answers. Green means OK, red "LTA", and blue that no answer to the question has been obtained. Irrelevant questions are crossed out. The analysis is complete when all elements have been covered.

Safety management and organization review technique (SMORT)

SMORT is a method inspired by MORT and is designed for simplified applications (Kjellén et al., 1987). It is based on the suppositions that the company in question has a safety policy, a plan for its implementation and a means for following up the results. The analysis is conducted on four levels:

1. On level 1 the courses of either accidents that have occurred or those that are conceivable are analyzed. The energy and deviation models are employed.

2. On level 2 production conditions which may explain why some problems have not been discovered and corrected are investigated.

3. On level 3 safety resources for the maintenance of safety in the design of new installations are studied.

4. On level 4 the internal information systems for safety, company safety policy and implementation at company level are examined

Check lists with relevant comments are available to help conduct the analysis at the various levels. The guide also contains a description of how an analysis is planned and implemented.

Whereas MORT is in principle a type of tree analysis, SMORT represents a more "linear" approach, i.e. the analyst works from the accident to the management function. Moreover, it is recommended that SMORT users thoroughly summarize results on one level before moving to the next. A further difference is that the assessment level "Less Than Adequate" is removed. Instead of making formal assessments, discussion takes place in order to provide answers to the various questions raised.

Some comments

The methods permit a large number of problems to be identified. Johnson (1980) mentions that 5 MORT studies of serious accidents led to the identification of 197 problems, i.e. about 38 problems per study. Johnson describes the method as simple. It is extensive, but each element is easy to understand. However, many perceive the method as impractible, perhaps because there are so many different items to keep track of.

Johnson suggests that the analysis of an accident can be conducted in one or a few days. However, experiences from Finnish applications of MORT to maintenance work (Ruuhilehto, K., 1991, pers. comm.) are that the analyst will require between two and eight weeks.

MORT and SMORT make use of good and penetrating questions. But both are based on an ideal model of organizations. Where the actual deviates from the ideal, there can be far too many negative answers, which analysts who lack a great deal of experience can find difficult to handle.

9.3 Brief descriptions of some other methods

Technique for human error rate prediction (THERP)

THERP is probably the most widely used and best known of the methods for analyzing and quantifying probabilities of human error. There is a handbook in which the method is extensively described (Swain & Guttman, 1983) and descriptions are also contained in other publications (e.g. Bell & Swain, 1983). The method has been developed steadily over a number of years.

The main stages of the procedure are:

1. Identification of system functions that are sensitive to human error.

2. Analysis of the job tasks that relate to the sensitive functions.

3. Estimation of error probabilities.

4. Estimation of the effects of human errors

5. When applied at the design stage, utilization of the results for system changes. These changes then need to be assessed further.

The handbook also contains tables with estimates of error probabilities for different types of errors. These probabilities may be affected by performance shaping factors, meaning that the analyst makes adjustments to their values in the light of the quality of the man-machine interface, experience of the individual operator, etc.

Action error analysis

This method provides another example of how procedures involved in the operation of technical systems can be analyzed. Its aim is to identify steps that are especially prone to human error and assess the consequences of such errors. The method is best suited for use in installations where there are well-defined procedures, e.g. in certain processing industries. If there are no well-established routines, it is difficult to find a basis on which the analysis can be conducted.

The method is simpler than THERP and has similarities with some of the forms of analysis described in previous chapters. Taylor (1979) has provided a description of the method, which is also known as the Action error method. The stages of the procedure are:

1. Making a list of the steps in the operational procedure. The list specifies the effect of different actions on the installation. It must be detailed, containing items such as "Press Button A" or "Turn Valve B".

2. Identification of possible errors for each step, using the check list of errors below.

3. Assessment of the consequences of the errors.

4. Investigation of conceivable causes of important errors.

5. Analysis of possible actions designed to gain control over the process.

Various conceivable **types of errors** include:

1. Actions not taken.

2. Actions taken in the wrong order.

3. Erroneous actions.

4. Actions applied to the wrong object.

5. Actions taken too late or too early.

6. Too many or too few actions taken.

7. Actions with an effect in the wrong direction.

8. Actions with an effect of the wrong magnitude.

9. Decision failures in relation to the actions taken.

Failure mode and effects analysis (FMEA)

FMEA is a well established method which has been utilized since the end of the 1950s. The method is well documented and several descriptions of its use are available (e.g. Hammer, 1972). There is also a standard description (IEC, 1985). This, however, has the character of a users' manual, providing guidance rather than being an attempt to standardize applications.

The method is employed for analyses of technical systems. FMEA can be used at different system levels, from individual components to larger function blocks. In simple terms, it is designed to answer the questions: "How can the unit fail?" and "What happens then?".

There are variations in the application of the method and the complexity of the systems analyzed. This means that details of the analytical procedure will also vary. The main stages in an analysis are as follows:

1. The system is divided up into different units in the form of a block diagram.

2. Failure modes are identified for the various units.

3. Conceivable causes, consequences and the significance of failure are assessed for each failure mode.

4. An investigation is made into how the failure can be detected.

5. Recommendations for suitable control measures are made.

There is a variant of the method called Failure mode, effects and criticality analysis (FMECA) which stresses the criticality concept. When using FMEA or FMECA a large number of possible failure modes will be discovered. It is necessary to set priorities between these, weighing in both the probability of occurrence and the seriousness of the effects. This can be accomplished in various ways, such as by dividing failures that have occurred into classes.

The IEC (1985) provides an example of a criticality scale, where the most serious level is: *"Any event which could potentially cause the loss of primary system function(s) resulting in significant damage to the system or its environment, and/or cause the loss of life or limb."*

It is best to use a special record sheet for the analysis. The IEC (1985) provides a version with 12 columns. Table headings may include:

— Identification — component designation, function.

— Failure mode.

— Failure cause.

— Failure effect.

— Failure detection.

— Possible action.

— Probability and/or criticality level.

A detailed analysis may be extensive. A system can contain a large number of components, and a component can fail in many different ways. A relay, for example, may have 21 different failure modes (Taylor, 1979). In the standard description (IEC, 1985) 33 generic failure modes are listed.

Event tree analysis

In a sense, an event tree is the opposite of a fault tree. A fault tree is constructed by selecting a top event and then working downwards. An event tree starts with an initiating event, e.g. a pump that ceases to operate, and then describes the consequences of this. Short descriptions are available, e.g. those of Lees (1980) and Taylor (1979).

The principle can be explained by an example. Let us consider premises where there is a container for poisonous gas and where a person is present. The container might leak, and so there is a gas detector. In the case of a leak, an alarm bell sounds and the person should then rush out of the premises. All leaks will not necessarily lead to gas being present in the workplace.

Every part of this sequence contains the possibility of success or failure. Failure can be due to the gas not reaching the detector. The alarm may fail to go off, or the person might not manage to get out. Figure 9.1 illustrates an event tree for a sequence which starts with a gas leak. There are two possible consequences. Either the person is not harmed or he/she is.

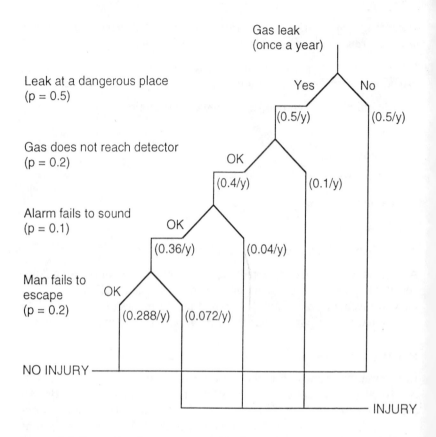

Figure 9.1 *Example of an event tree for the consequences of a gas leak*

An event tree provides opportunities for making probabilistic estimates. The initial event is expressed as a frequency (events per year). The other branch-off points are expressed as probabilities (the number of failures per trial or occasion of use).

Figure 9.1 provides an example of how an estimate can be made. For purposes of clarification, rather high frequency and failure probability values have been used. On the basis of these values, the frequency of injury to a person resulting from a gas leak is 0.2 times per year (0.1 + 0.04 + 0.072).

148

Cause-consequence diagrams

One technique, related to both Event tree and Fault tree analysis, involves Cause-consequence diagrams. The analysis starts with the definition of a critical event in the system. Possible causes of the event are then filled in, as with the construction of a fault tree. Consequences are also investigated, as when an event tree is constructed. Descriptions of the method have been provided by Nielsen (1971, 1974) and Taylor (1974, 1979)

One feature of the method is that alternative consequence paths can be handled. It can also be used as a basis for making probabilistic estimates.

Change analysis

Changes to a system can create new hazards or a deterioration in the control of hazards that are already known. Change analysis is designed to identify causes of increased risks arising from system changes. The method was originally designed for application to organizational systems (Kepner & Tregoe, 1965). Its aim is to identify the basic changes which give rise to a problem. Change analysis has been employed since the 1960s.

Descriptions that focus on accidents which have occurred are available in the literature (e.g. Bullock, 1976; Johnson, 1980). Application of the method involves comparison of the situation on the occasion of the accident with a non-accident situation. The most important steps in an analysis are:

1. Comparison of the accident and non-accident situations.

2. Noting all known differences.

3. Analysis of the differences, and assessment of the influence of these on the accident.

A special record sheet can be used for the analysis. On one axis there is a list of 25 factors which may be subject to change. On the other, there are headings for the description of current and previous situations, differences, and changes that have an effect.

This type of Change analysis is applied to accidents that have already occurred. Both planned and unforeseen changes are included. The method has similarities to deviation investigation. However, in Change analysis it is assumed, more or less implicitly, that the old system has an adequate level of safety. This is sometimes a weakness of the analysis.

9.4 Quick analyses

Even simple analyses are of value and provide information on existing hazards. In many situations there is no justification for a stricter form of safety analysis, or the opportunities for one are lacking. Such situations can include:

— Presence of major safety deficiencies. If it is already known that there are a large number of safety problems, no detailed analysis is needed for these to be identified. If the deficiencies are major, making improvements is a matter of urgency. The measures taken can then be developed gradually using stricter forms of analysis.

— Unclear picture of the hazards. It is not known whether a thorough investigation is justified.

— Lack of resources. A full analysis cannot be conducted because of lack of people or time.

— Absence of documentation on the existing system or planned changes. There is not sufficient information available for a proper analysis to be conducted.

A rough analysis is a compromise between safety analysis and the unsystematic discussion of possible hazards. Such an analysis tends to have the following features:

— It is quicker to conduct than a normal safety analysis.

— It is not fully systematic. This might mean that only certain aspects of the system are considered, or that only specific types of hazards are investigated. The aim, however, should be to cover the system as a whole.

Several of the methods already referred to can be used in a "rough" or "quick" manner to cut down the time taken by an analysis. A short summary of a variety of approaches is provided below. Sometimes the methods will overlap, so that one approach can contain some of the elements in another:

1. Use of check lists — based on summaries of known problems.

2. Comparisons with directives.

3. Comparisons with similar installations.

4. Inventories of documented hazards.

5. Inventories of known hazards.

6. Preliminary hazard analysis.

7. "What-if."

8. Quick energy analysis.

9. Quick deviation analysis.

1) Use of check lists

A check list of problems may have been obtained from a general problem inventory. The installation in question is checked against the list to see if these problems exist. The quality and utility of the analysis will be improved if the list is adapted for a particular type of installation or situation, e.g. for the planning stage. Check lists have been developed for a variety of situations and specific industrial sectors.

2) Comparisons with directives

Directives issued by the authorities can be treated as check lists. They represent a summary of knowledge obtained over a long period of time. Also, of course, checks can be made on whether the directives are being followed. However, such a procedure will seldom be as thorough as a safety analysis.

3) Comparisons with similar installations

If there are similar installations where hazards have been more thoroughly investigated, an analysis can be based on the knowledge thereby obtained. The aim of the analysis is to check whether the same hazards exist at the installation to be studied. This form of analysis may be appropriate when a new installation is planned or when changes are made to an existing installation.

4) Inventories of documented hazards

In the case of installations that are in operation, there is generally documentary material available on injuries and damage, e.g. accident investigation reports. An inventory can then be obtained by preparing a structured summary of these.

5) Inventories of known hazards

The purpose of taking an "inventory of known hazards" is to summarize the hazards that are known to employees and can be reported spontaneously. Such an inventory can be taken in many ways. One way is to arrange a meeting of a suitably-composed team, which then focuses on the most serious accident hazards in various parts of the installation. Good experiences have been had of so-called "brain-storming" techniques. For example, one way of beginning is to have all participants write down the ten hazards that they consider to be most serious.

6) Preliminary hazard analysis

Hammer (1980) has proposed the use of a method which has acquired the name "Preliminary hazard analysis". The first step is to summarize known problems. The function of the system and its surrounding conditions are then summarized, and an attempt is made to identify the more serious hazards, etc. When applied in full, the method is fairly extensive.

7) "What-if"

"What-if" analysis is a popular technique employed in many parts of processing industry. It is not a specific method with a standardized application, but varies according to the user. The basic idea is to pose questions such as:

— *What* happens *if* Pump A fails?

— *What* happens *if* there is an interruption to the electrical power supply?

— *What* happens *if* the operator opens Valve B instead of Valve A?

If the right questions are posed to a skilled team, good results can be obtained. But success using this method is very much dependent on the extent to which the approach is systematic and on the skills of the users.

152

8) Rough energy analysis

An Energy analysis can easily be simplified to provide a simple hazard survey. Simplification involves dividing the system into large sections and only considering energies that can lead to fatal or serious injuries.

9) Rough deviation analysis

Deviation analysis can also be simplified. This means that the division into functions (the structuring) is done quite crudely. Only deviations with relatively major consequences and important planned changes are considered.

10 Methodological overview

10.1 Comparison between methods

There are a large number of different methods for safety analysis. This book has referred to around 20, including quick analyses. Two compilations (Clemens, 1982; SCRATCH, 1984) made at the beginning of the 1980s jointly cover 37 methods, five of which were taken up by both studies. The overlap was small, which suggests that the real number was considerably higher. Since then, one can be sure that a number of further methods have been developed. The previous chapters have described some methods for safety analysis with a varying degree of thoroughness. Table 10.1 provides a summary of 10 methods, which jointly represent a variety of approaches and ideas.

Many separate issues concerning aspects of safety analysis can be raised by comparing methods. For example, the SCRATCH project (1984) considers 12 different issues, while Clemens (1982) takes up four. In terms of the basis for classification, practice varies within the literature (e.g. Suokas, 1985). The evaluations below are primarily concerned with the application of the methods specifically to the field of occupational accidents.

There are several difficulties involved in making general evaluations. One reason for this is that the results of a safety analysis are determined to a great extent by who has conducted the analysis. Adoption of a particular method will not in itself guarantee good results. Another reason is the great variation in areas of application and differences between users.

Different aspects of safety analysis

Through making a comparison between methods a number of different themes will be discussed:

— Modelling of the system (of reality).

— Structuring of the system, i.e. how the analytical model is applied.

— Use of models of how accidents occur.

— Identification, i.e. what is identified and how.

Table 10.1 *Summaries of 10 methods for safety analysis*

Method	Summary
1. Deviation analysis	Identifies deviations from the planned and normal production process which can lead to hazards.
2. Hazard and operability studies (HAZOP)	Identifies deviations from intended design which can lead to hazards.
3. Action error analysis	Identifies departures from specified job procedure which can lead to hazards.
4. Failure mode and effects analysis (FMEA)	Identifies failures of components (or modules) which can lead to hazards.
5. Energy analysis	Identifies energies which can harm human beings.
6. Job safety analysis	Identifies hazards in job procedures.
7. Fault tree analysis	Analyzes causes and shows logical connections which can lead to a specified undesired event.
8. Event tree analysis	Analyzes alternative consequences of a certain hazardous event.
9. Management oversight and risk tree (MORT)	Compares a model of organizational requirements with an existing organization.
10. Change analysis	Establishes the causes of problems through comparison with problem-free situations.

— Qualitative and quantitative analysis.

— The systematic approach to safety measures, which is directly linked to the method.

— Analytical procedure.

— Time to analyse.

— Difficulty of the method.

155

MODELLING AND STRUCTURING THE OBJECT OF STUDY

How an object is treated by the different methods and how its different parts are included in the analysis are roughly summarized in table 10.2. To be able to describe an installation some sort of analytical model of the system is required. Most analytical methods contain a more-or-less explicit model of the reality to be analyzed. Under the heading "Aspects" in the table, it is specified whether a certain method focuses on technical and physical aspects of the system (T), on the individuals who work there (I), or on organization and management (O).

Two of the models, HAZOP and Energy analysis, are purely based on technical system properties. This also generally applies to FMEA and to Fault tree analysis, but both these methods can have a wider area of application. In the author's view, three of the methods involve the adoption of an overall perspective, requiring consideration of techniques, individuals and the organization. These are Deviation analysis, MORT and Change analysis.

The first six methods are based on consideration of the object, followed by the construction of a "model" on the basis of this. Methods 1, 2 and 5, and to a certain extent methods 6 and 9, are designed to cover the entire object. Methods 3 and 4 focus on a certain part of an installation, either a specific job procedure or a technical part of the system. Methods 7, 8 and 10 are not based directly on the object itself, but show how a certain event or problem is related to it.

Structuring has been divided up into four classes according to how the structure relates to the object. Methods 1 and 6 involve major structuring which needs to be well thought through. For some of the methods, a fairly uniform structuring procedure can be adopted, while for others a structure can be created in a variety of different ways.

MODELS OF ACCIDENTS

The various methods also contain a more or less explicitly expressed model of how accidents occur. In most cases, this is described in the chapters above. Under the heading "Aspects" in table 10.2 there is a list of the features of the object which are considered relevant. In a sense, this specifies whether technical, human or organizational factors are regarded as lying behind the occurrence of an accident.

Table 10.2 *Structuring of the object of study using different methods*

Method	Structuring	Type	Aspects
1. Deviation analysis	On the basis of activities, e.g. the production flow or job procedure.	3	T, I, O
2. HAZOP	On the basis of physical properties, e.g. pipes and tanks.	2	T
3. Action error analysis	On the basis of a detailed description of the phases of work of the operator.	1	I (T, O)
4. FMEA	On the basis of technical components or modules.	1	T
5. Energy analysis	On the basis of volumes, which jointly cover the entire object.	2	T
6. Job safety analysis	On the basis of an individual's job tasks.	2	T, I
7. Fault tree analysis	Not on the basis of the object, but in relation to the top event.	4	T (I)
8. Event tree analysis	Not on the basis of the object, but in relation to the initial event.	4	T (I)
9. MORT	Not on the basis of the object, but in relation to a general model of organizations.	4	T, I, O
10. Change analysis	Not on the basis of the object.	4	T, I, O

Key:
Type of structuring
1. Determined by the object description
2. Needs development, but partially pre-determined.
3. To be developed with care
4. No special structuring required

Aspects of the object
T Technical
I Individual actions
O Organization

An energy model clearly lies behind three of the methods, i.e. Energy analysis, Job safety analysis and **MORT**. Many of the other methods also include an explanation of accidents that is couched in energy terms.

IDENTIFICATION OF HAZARDS AND DEVIATIONS

A further feature of each method is how hazards are identified. All the methods involve consideration of deviations in one way or another. This aspect of the methods is summarized in table 10.3.

What can be conceived of as "deviating" is determined by which aspects of the object are considered (the final column in table 10.2). The next issue is how deviations are discovered. For several of the methods check lists are available as aids for identification. These are of two types:

1. Check lists of types of deviations.

2. Check lists of what can deviate.

Check lists of types of deviations

Several of the methods involve the use of a check list of types of deviations. If the analysis is limited to specific properties of the object, a list that provides detailed and thorough coverage can be prepared (methods 3, 4 and 9). In HAZOP, attention is mainly restricted to physical parameters.

Check lists of what can deviate

In two of the methods, emphasis is placed on what can deviate. The check list in Deviation analysis specifies the system properties that might deviate. The MORT list refers in detail to the characteristics under study, and how each and every one of these might deviate.

Several of the methods utilize a "binary" classification, i.e. the deviation is defined and is assumed to be capable of either occurring or not occurring. Binary classification is most evident in the cases of methods 7, 8 and 9, but also applies to methods 3 and 4. Using the other methods, deviations are not defined and treated in such a manner.

QUALITATIVE AND QUANTITATIVE ANALYSES

One possible division is into qualitative and quantitative methods. Quantitative in this context means probabilistic, i.e. probabilities can be calculated. Both Fault tree and Event tree analysis belong to this latter

158

Table 10.3 *Deviations and their identification using different methods, plus the availability of rules for safety measures*

Method	Deviation from:	Check list	Rules
1. Deviation analysis	Planned and normal production process.	1 & 2	Yes
2. HAZOP	Intended design of system function.	1	-
3. Action error analysis	Specified phase in job procedure.	1	-
4. FMEA	Technical function of component or module.	-	-
5. Energy analysis	Control of energy or function of barrier (triggering-off an energy flow).	-	Yes
6. Job safety analysis	(Safe way of working.)	-	In part
7. Fault tree analysis	Normal system function.	-	-
8. Event tree analysis	Function for restoring the system to a safe state.	-	-
9. MORT	Organizational model (specified criteria for "Less Than Adequate").	1 & 2	-
10. Change analysis	Accident-free system.	1	-

Key:
Type of check list
1. For identifying types of deviations.
2. For discovering what might deviate.

category, but they can be used without calculating probabilities. In some cases, probability values can be generated by FMEA. The other methods are qualitative by nature.

THE SYSTEMATIC APPROACH TO SAFETY MEASURES

In principle, all the methods provide a basis for the generation of safety measures. Two of the method descriptions (1 and 5) provide a set of rules which provide assistance in generating ideas for safety proposals.

ANALYTICAL PROCEDURE

Analytical procedures can be classified in a variety of ways (e.g. SCRATCH, 1984; Suokas, 1985). One possibility is to base the classification on the aims of the analysis and create a number of main categories:

1. *Hazard identification.* The aim is to discover which hazards (undesired events) might occur. This means that efforts are made to identify conditions that might lead to certain kinds of hazards.

2. *Accident modelling and accident event description.* How an accident might conceivably occur and what its consequences might be are described in formal terms.

3. *Structured comparison.* A comparison is made with some type of "normal system".

The aim of the first six methods is hazard identification, through the adoption of a specific analytical procedure. The analysis is based on the object in question, which is divided into parts, each of which is then treated separately.

Fault tree and Event tree analysis are examples of methods that involve accident modelling. Such a procedure is more iterative by nature. This means that steps in the analysis cannot be taken and completed one at a time.

The third category covers examples of structured comparison. MORT and Change analysis belong in this category. Using MORT, a comparison is made with a model organization, while in the case of Change analysis a non-accident situation is compared with one where an accident occurs. MORT can be applied in different ways, but these applications are usually guided by a specific analytical procedure.

10.2 Some advice on choice of method

It is difficult to give general advice on selection of method, and most general guidelines should be treated with caution. Nevertheless, although the choice will depend on a wide range of factors, some of the following might be taken into account:

1. Accuracy requirements.

2. Nature of the object.

3. Time needed.

4. Skills available and difficulty of the method.

Accuracy requirements

If only an overview of the hazards is required and the level of ambition is low, some form of "quick" analysis can be used. An intermediate level, which lies somewhere between a rough overview and an in-depth analysis, can be obtained using Energy analysis or Job safety analysis. Deviation analysis and HAZOP can be employed with varying degrees of detail and accuracy. For detailed analysis of a specific system, Fault tree analysis or FMEA can be selected.

Information on the object

For methods 2, 3, 4, 7, 8 and 9, a detailed description of the object is usually needed, although it is not always essential for methods 2 and 8. Methods 1, 5 and 6 do not require such detailed knowledge. It should also be remembered that systems vary with respect to the extent that activities are regulated and the freedom offered to the operators. Methods for which detailed information is required work less well when they are applied to relatively unstructured activities.

Time needed

The time required to conduct an analysis depends on a large number of factors. These may include familiarity with the form of analysis, the size of the object, the extent of the availability of information and accuracy requirements. Table 10.4 provides a rough picture of the time required to conduct each type of analysis. It is assumed that the object is of limited size and complexity. Given that the methods have different areas of application, it should be noted that the systems are not fully comparable.

161

The quickest methods are Energy analysis and Job safety analysis. Also, a Deviation analysis can usually be conducted in a relatively short time, if the level of ambition is not too high. Some examples of specific analyses, showing the time taken, are presented in chapter 15.

Table 10.4 *Time and documentary material requirements for the different methods*

Method	Time needed for analysis	Comments	Information needed on the object
1. Deviation analysis	1d (4h)	Varies considerably according to degree of detail.	2-3
2. HAZOP	1 - 2w (1d)	15 min. per pipeline. 2 h per main unit.	1-2
3. Action error analysis	-	Information not available on time required.	1
4. FMEA	1d - 2w	For example, 30 min. per component in a relay system.	1
5. Energy analysis	4h (1h)	-	2-3
6. Job safety analysis	6h (2h)	5-10 min. per work phase.	2
7. Fault tree analysis	1d - 2w	-	1-2
8. Event tree analysis	1d - 2w	-	1-2
9. MORT	1w - 2w	Varies considerably according to the level of ambition.	1
10. Change analysis	-	Information not available on time required.	2 (1)

Notes:
Time needed: The time required for structuring before analysis takes place is given in brackets.
Time units: min. = minute, h = hour, d = day, w = week.
Information needed on the object: 1 = detailed information, 2 = comprehensive information, but not detailed, 3 = general description or oral report.

Skills required and difficulty of the methods

The methods vary in terms of the skills they require and the time they take to learn. Again, it is difficult to provide general guidelines. The effort needed will depend on previous knowledge, how easy it is to get involved in how a technical system operates, etc. But there are, of course, differences between the methods themselves.

People attending courses on safety analysis have been requested to give their views on how difficult some of the methods are to use. This applies to methods 1, 2, 5, 6 and 7. The students' experience of each method consisted of theoretical instruction of about one hour and exercises lasting half a day.

Energy analysis and Job safety analysis were considered the easiest, followed by Deviation analysis and HAZOP. The participants regarded Fault tree analysis and MORT as the most difficult.

11 Risk assessment

11.1 Introduction

As described in chapter 1, the literature on safety analysis contains a variety of terms, the meanings of which can vary. In this book, safety analysis is treated as having three principal components: hazard identification (identification of a source of risk), risk assessment and the generation of safety proposals. Risk assessment, with which this chapter is concerned, has two subcomponents: risk estimation (making estimates of probabilities and consequences) and risk evaluation (making an overall judgement on the importance of a risk, e.g. in terms of its acceptability and how it is perceived).

In most cases, risk assessment forms an important part of a safety analysis. The seriousness of an identified hazard needs to be evaluated. In some of the methods described, risk assessment constitutes a specific stage in the analytical procedure.

Usually, a list of identified sources of risk will have been obtained. Each hazard needs to be assessed to permit a judgement on whether safety measures of some type are required. This chapter focuses on types of risk assessments that are relatively simple and concrete by nature.

With a little simplification, risk assessments can be divided up into three main groups:

1. Informal assessments.

2. Qualitative assessments (of the risk or the need for safety measures).

3. Quantitative assessments, followed by evaluations of their results.

Informal assessments

In this context, an informal risk assessment is one which is not a planned part of the analytical procedure and not based on any specific documentation of risk.

In connection with the identification of hazards, a number of informal risk assessments are made. These may lead to a hazard not being included on the record sheet. This might be because the probability of the occurrence of the event seems low or that its consequences seem minor. Or, there may be other justifications, of a greater or lesser degree of validity. In practice, such omissions cannot be avoided, as the number of identified hazards may be large. On the other hand, it is not acceptable that hazards are dismissed on the grounds that they are "just part of the job", etc. Thus, an awareness of the problems involved in risk assessment is needed even at the hazard identification stage.

Qualitative assessments

A qualitative risk assessment is one which involves making a judgement on the seriousness of a hazard in the absence of quantitative measures of risk. Examples of such judgements are that a risk is acceptable (or not) or that safety measures should be taken (or not). Or, hazards can be placed in rank order of importance as, for example, in table 8.2.

Quantitative estimates

In some situations, the probability that a certain accident will occur and the size of its consequences can be calculated or estimated. The quantitative measure of risk obtained can then be utilized to judge whether or not the hazard is acceptable. This can be achieved by making a comparison with other hazards or by referring to some type of norm or guideline.

Difficulties

It is often difficult to assess accident risks. It cannot be assumed that all analysts will come to the same conclusion. Objective results, those that are independent of the assessor, are impossible to obtain. There is a subjective element to risk assessment, which stems from differences in attitudes and values.

Norms

If certain principles can be applied, the preconditions for making objective assessments are obviously more favourable. The problem, however, is that the availability of clear and unambiguous norms is the exception rather than the rule.

In principle, one basis on which accident risks can be assessed is offered by the directives issued by the authorities. These, however, are mainly general by nature and do not cover all types of hazards. In some situations and for certain types of equipment, fairly concrete information can be obtained on whether or not a risk is acceptable. But there are also many formulations of the type "Protection against injury shall be adequate" or "that risks should be As Low As Reasonably Achievable (ALARA)". To establish what is "adequate" or as "low as reasonable" is again a matter of judgement.

11.2 Consequence and probability

Consequence

The methods discussed so far are not generally intended for estimating the size of consequences. The method that comes closest is Energy analysis, where it is sometimes possible to establish whether an energy is large enough to cause a fatality, to lead to serious injury, etc. In such cases, "energy magnitudes" can be assessed. For example, the amount of a poisonous substance or the height at which a person is working can be compared with an energy value that a person might be expected to cope with unscathed.

There are a number of different estimation methods available for different types of gas emissions, and for events related to fires and explosions. These are primarily applied where accidents involving chemicals might occur. An account of this type of estimation falls beyond the scope of this book, and the reader is referred to the more specialized literature (e.g. DiNenno, 1988; Centre for Chemical Process Safety, 1989). There is a wide range of computer programs available to aid this kind of estimation (Marnicio et al., 1991).

Probability of occurrence

Fault tree, Event tree and Failure mode and effects analysis are examples of methods where probabilities for the occurrence of events can be estimated. In general, these are applied to technical systems where the consequences of a certain event are rather clear. Such estimates are often associated with relatively major uncertainties. Despite the problems they involve, such calculations can be of benefit in providing the basis for setting priorities and for choosing between alternative control measures.

Combined assessment

Probabilities and consequences can be combined to provide different types of risk measures. Examples of a variety of measures that might be utilized for a single installation include:

— Probability that there is a fatality during time period t.

— Probability that there is a fatality within a specified distance from a source of risk (isorisk curves) during time period t.

— The mean number of fatalities over, for example, a period of 100 years.

"Common" accidents

When applied to common occupational accidents, a safety analysis could hypothetically result in the production of a summary account of consequences and probabilities for different types of injury events. It should then be possible to use the results to make a theoretical estimate of the number of accidents that will occur at a certain installation and the days absent to which they will give rise.

However, there are a number of problems involved in predicting the future accident picture. The problems are so great that there is seldom any point in trying to make a quantitative estimate. Examples of these problems include:

1. *Statistical uncertainty.* Even if the calculations are of high quality, the stochastic error in predicting the outcome at a particular install-ation can be large. If the analysis is applied to a mass-produced piece of machinery, an estimate of this sort may be more meaningful.

2. *Estimation uncertainties.* These will also be large. As a guess, the uncertainty factor should be about 10. The tolerances for accident risks are generally less than this. It can be assumed that there is an average frequency of 40 industrial accidents per million hours worked. A frequency of 4 will denote a safe workplace, while 400 means that the risk is very high. (The figure 4 is obtained by

dividing the average frequency by the uncertainty factor, the figure 400 by multiplying the two values.)

3. *Information on probabilities*. Data are generally lacking for a variety of types of relevant basic events.

4. *Consequences*. The consequences of a certain event can vary considerably. For example, a fall from a height of one metre may lead to very serious injuries, but a person might also escape unharmed.

5. *Behavioural adjustment*. People take account of occupational hazards to a greater or lesser extent. There is no unequivocal relationship between the presence of physical hazards and the occurrence of accidents.

6. *Time required*. Working on the calculations is time consuming.

One way of assessing identified hazards is to classify them according to the consequences of hazardous events occurring and the probabilities that they will do so. Two examples are given below.

Table 11.1 *Example of the classification of consequences*

Code	Category
0	Not harmful or trivial
1	Short period of sick leave
2	Long period of sick leave
3	Disablement
4	Fatality
5	Several fatalities, major disaster

Table 11.2 *Example of the classification of probabilities*

Code	Category	Probability
0	Highly unlikely, impossible	1 in a 1 000 years
1	Unlikely	1 in a 100 years
2	Rather unlikely	1 in 10 years
3	Rather likely	Once a year
4	Likely	Once a month
5	Very likely	Once a day

Table 11.2 gives examples of probability values. These have a wide range — from 1 to around 10^{-6} times per day. The scale is intended to include types of disturbances that may be quite common, but also covers events that occur more rarely, such as accidents and catastrophes.

It is especially difficult to make a probability estimate for events that only rarely occur. As a guide, it can be said that about one in every twenty industrial workers receives an occupational injury each year, i.e. a code of "2" on the scale given in table 11.2. This represents the sum of all the different risks at work.

In theory, probability and consequence can be combined into a single measure of risk. Usually, the two values are multiplied. But, an alternative assumption is that the factors are logarithmic, in which case the summary measure of risk is obtained by adding the two values. In making a strict estimate, varying probabilities for different consequences should be taken into account.

Using estimated or calculated values in risk assessment is not without problems, but it does permit comparisons to be made between different identified hazards. Experience of such applications has been favourable. Their primary advantage is that it forces hazards to be discussed in a fairly systematic manner.

11.3 Practical risk evaluation

Despite the difficulties involved, risks have to be judged and decisions made in relation to them. One way is to utilize a summary measure, which is then applied to the identified hazards. An example is provided in table 11.3.

In principle, use of the scale permits risks to be classified into two main

Table 11.3 Example of a summary risk scale

Code	Description
0	Negligible risk
I	Acceptable risk, no safety measure required
II	Safety measure recommended
III	Safety measure essential

categories: those that are "acceptable" (O, I) and those that are "not acceptable" (II, III). In concrete terms, "not acceptable" means that a safety measure is required. Examples of the bases on which such a conclusion could be reached are as follows:

1. Breach of a directives issued by the authorities.

2. Breach of company regulations.

3. Installation with poor accident record.

4. An energy high enough to lead to serious injury.

5. Low tolerance of the system for errors that can trigger the hazardous event. This may mean that a high level of attentiveness is required from personnel for an accident to be avoided.

6. Suitable solution available.

The directives of the authorities sometimes provide a reliable basis for assessment. On occasions, a directive is sufficiently precise for it to be established directly that something must be done about a particular hazard. In such cases, it is appropriate to make a note on the analysis record sheet which refers directly to the section of the directive in question. This should also be done where the company's own regulations apply.

Evaluation in teams

There are advantages in making risk evaluations in a suitably-composed team. Where agreement is not reached, this can then be noted on the record sheet. However, agreement is not absolutely necessary. A more definite position can be adopted when decisions on control measures are to be taken.

The author's own experiences of team work are that it gives rise to relatively few divergent assessments. Rather, the differences in judgement are less than might be expected in the light of the different values that people hold and the various positions they occupy within the company.

A further justification for the team approach is that an excessive amount of time tends not to be devoted to risk assessment, when this time could be better spent on producing proposals for safety measures. In the course of discussing such proposals, an increased understanding of the hazard in question can be obtained, and this may itself have an influence on how risks are assessed.

Definition of criterion

The criteria for evaluation ought to be discussed before embarking on the risk evaluation itself. Experiences of previous accidents in the same or similar installations and reports on these can be of value.

Other perspectives

It has been more-or-less implicitly assumed that the assessments in question have concerned the risk of injuries to human beings. It may be of advantage to provide a slightly wider perspective on risk assessment. For example, the risk of a disturbance to production or a deterioration in product quality can be included. If a disturbance might have several different consequences, it is a good idea to take these into account in the assessment. This issue is further discussed in section 13.5.

It is impossible to get away from financial considerations, i.e. what safety measures might cost. Nevertheless, in making a risk assessment, an attempt should be made to ignore these factors. In principle, it is only consequences and probabilities that count. The appropriate time to take up financial issues is when discussions take place on which measures should be implemented, how available resources should be utilized, and which priorities should be set. See also chapter 12 and section 13.4.

12 Safety analysis — planning and implementation

12.1 Introduction

The implementation of a safety analysis involves more than the application of one or several methods. Conscious planning is required for good results to be obtained. When a large installation is to be analyzed and many people are affected, planning is of extra importance. If the analysis is limited in scope, planning can be made much simpler. Nevertheless, some of the views expressed in this chapter will still be useful.

This chapter takes up a variety of aspects of the planning, implementation and utilization of safety analyses. The chapter that follows on sources of error and the analytical problems that may arise supplements this description. There is a certain amount of literature available which takes up different aspects of the planning of analyses (e.g. CISHC, 1977; Kjellén et al., 1983; SCRATCH, 1984).

There are an increasing number of computer programs available as aids for conducting analyses (e.g. Marnicio et al., 1991; Heino et al., 1992), but only few of these are applicable to occupational accidents. Even when computer facilities are utilized, it is important to follow the analytical procedure and not allow the computer program to "take over".

Table 12.1 Tables related to safety analysis planning

Table content	Table
Aims of an analysis	12.2
Types of objects analyzed	12.3
Situations during system life cycle	12.4
Categories that may be affected by an analysis	12.5
Arguments for safety analysis	12.6
Arguments against safety analysis	12.7
Summary of stages in safety analysis	12.8
Example of a plan for a safety analysis	12.9
Examples of sources of information	12.10
Time requirements for some methods	10.4
Costs and benefits of safety analysis at company level	13.2
Examples of cost-benefit analysis	13.4

This book contains a number of tables that can be utilized as planning check lists. A list of these is provided in table 12.1

12.2 Preconditions and preparations

Before a decision is made to conduct an analysis, some basic preconditions for the success of the analysis should be investigated. Summaries of a number of different aspects are provided in this section. These are as follows:

1. Aim of the analysis.

2. Type of object to be analyzed.

3. Stage in the life cycle of the object.

4. Interested parties.

5. Arguments for and against conducting an analysis.

Table 12.2 Examples of the aims of a safety analysis.

IDENTIFICATION AND ASSESSMENT Identification of accident hazards. Identification of causes of breakdowns of equipment. Identification of causes of interruptions to production. Risk estimation and evaluation (risks and consequences of certain accidents). Survey of the capacity of the organization to handle safety matters. Basis for emergency planning. Assessment of product safety. Evaluation of safety in relation to certain criteria, e.g. directives of the authorities.
FINDING MEASURES TO INCREASE SAFETY Improvements to machines and layout. Improvements to job procedures and job instructions. General improvements to organizational routines. Provision of assistance to operators at work.
MISCELLANEOUS Meeting the requirements of the authorities. Initiation of a process to promote safety consciousness within the company. Improvements to and systematization of safety work. Provision of basis for decisions to be taken, e.g. for setting priorities or making investments. Learning how to utilize safety analysis at company level.

Aim of the analysis

Table 12.2 provides examples of the various aims that a safety analysis may have. Of course, having one objective does not exclude consideration of others. On the contrary, it may be of advantage that several objectives are achieved at the same time (see also sect. 13.5).

Type of object to be analyzed

The nature of the object will determine the choice of approach and method. Table 12.3 provides examples of different types of objects, while stages in the life cycle of a system are shown in table 12.4.

Table 12.3 *Examples of types of objects to be analyzed*

Type of system	Comments
Entire installation	Overview required at the start of the analysis.
Type of machine	Standard machine.
Specific machine	Machine tailor-made for the user.
Part of the workplace	Relevant activities and equipment.
Transportation system	Affects technical equipment and routines.
Specific type of work	E.g. carrying out a repair.
Organizational routines	E.g. maintenance routines or planning procedures

Interested parties

People in a variety of different positions will usually be directly affected by the analysis and the changes to which it may give rise. Some may be in possession of important information or have views on recommendations and decisions. Table 12.5 provides examples of the various categories of people that may be affected at different stages of the analysis.

Arguments for and against a safety analysis

The first step to be taken when planning a safety analysis is to decide whether an analysis should be conducted. There may be arguments both for and against. Some examples of these are shown in tables 12.6 and 12.7.

There may be reasonable grounds for all the arguments (both for and against) contained in the tables. For this reason, they need to be evaluated in the light of the concrete situation. A number of additional viewpoints

Table 12.4 *Examples of situations arising during the system life cycle*

PRE-OPERATION
Conceptualization of the system
Pre-planning
Planning
Procurement
Construction
Inspection on delivery
System start

IN OPERATION
Normal operation
Preventive maintenance
Repair
Changes to:
— Technical equipment
— Routines
— Staffing
— Computer programs
— Products

DISPOSAL
Dismounting the system
Clean-up of site

Table 12.5 *Examples of categories of people that may be affected by a safety analysis*

WITHIN THE COMPANY
Company executives
Production staff and job supervisors
Personnel at the installation
Maintenance staff
Design staff
Buying staff
Safety committee members
Safety representatives
Trade union representatives

OUTSIDE THE COMPANY
Authorities responsible for safety matters
Suppliers of equipment
Consultants, e.g. those involved in the design of a production system
Customers, e.g. when changes are made to a product
Occupational Health Services
Insurance companies
The media

Table 12.6 *Arguments for conducting a safety analysis*

GENERAL
1. Demands have been made by the authorities, or legal requirements need to be satisfied.
2. Safety management needs to be improved.
3. There is a need for an overall picture of the changes required.
4. Systematic analysis would make it easier for production personnel to effectively contribute the fruits of their experience.
5. Improved productivity would result from greater operational reliability.
6. A good basis for training and better job instructions for maintenance operations workers would be provided.
7. Profitability gains would result from better workstations and fewer problems.
8. An analysis would provide a good basis for the dissemination of information on operational safety and hazards - to employees, authorities and the general public.
9. Insurance premiums could be reduced.

AT THE PLANNING STAGE
10. There appear to be serious accident risks at the new installation, e.g. high energies.
11. There are safety problems at similar installations.
12. The consequences of failure may be serious.
13. There is an opportunity to prevent various types of problems and eliminate certain hazards.
14. It would be cheaper to eliminate hazards and solve problems before they have been built into the system.
15. A better basis on which to list specifications to suppliers could be obtained.
16. A better basis for checking on equipment supplied could be obtained.
17. System start problems could be reduced and a shorter start-up period could be obtained.
18. Understanding of the new installation could be facilitated, as a result of going through different system functions systematically.

FOR EXISTING INSTALLATIONS
19. There is a serious accident problem and many accidents occur.
20. Attempts to reduce the number and severity of accidents have failed.
21. There are many disturbances to production and correction systems function irrationally.

on the benefits of safety analyses are provided in the chapters on safety management and its implementation (chapts 14 and 15).

12.3 Analytical procedure

The discussion of planning below is principally based on the model of safety analysis provided in figure 3.1. Table 12.8 offers a more extensive

Table 12.7 *Arguments against conducting a safety analysis*

GENERAL 1. There is no safety problem. 2. Hazards are already known, only funds being required for the necessary safety measures to be taken. 3. There would not be any accidents, if only personnel would follow existing job instructions. 4. Benefits of an analysis are hard to assess. 5. An analysis would be expensive and disruptive to work. 6. An analysis might lead to demands for major and expensive improvements. **AT THE PLANNING STAGE** 7. Project is far too advanced for a safety analysis to have any effect. 8. Project is at far too early a stage (inadequate knowledge of the form that the installation will finally take). 9. Knowledge of the installation is too poor for an analysis to be undertaken. 10. Responsibility that the machine meets all requirements and is sufficiently safe lies entirely with the supplier. 11. Influence on the supplier to make improvements cannot be exerted (lack of willingness on the part of the supplier). **FOR EXISTING INSTALLATIONS** 12. There is no safety problem, accidents either not occurring or being trivial by nature. 13. The Labour Inspectorate has visited the installation and made no complaints. 14. Production facilities may soon be changed.

picture of the analytical procedure. It can be used as a check list before planning an analysis.

Initiation of the analysis

The first step is to make a proposal that an analysis should be undertaken. Someone must take the initiative, and a decision must then be taken. Examples of arguments for and against conducting an analysis are shown in tables 12.6 and 12.7.

Study team and/or management group

It is often preferable to conduct a safety analysis in a team. An extensive analysis can affect a large number of people. A team (or several) can share the work of data collection, analysis, etc. In the case of a major analysis, it may be a good idea to appoint a management group. It will

Table 12.8 *Summary of stages of procedure in a safety analysis*

Stage	Comments
BEFORE THE ANALYSIS Initiation	An analysis requires an initiative to be taken, and needs to be rooted at company level, etc.
PLANNING Appointment of study team	A study team and/or management group may be needed.
Definition of aims	Objectives and subobjectives are specified.
Setting object limits	How large a part of the object and which activities are to be covered?
Specification of assumptions	What assumptions should be made about the object?
Planning	Work procedure, time and financial resources required. Coordination with other activities, e.g. other projects and investigations.
Follow-up	Readiness to make adjustments and follow up planning results.
Information program	Those affected by the analysis should receive information.
INFORMATION On the system	What information is required? Where can it be found?
On the "problem"	As above. The information might concern accidents, disturbances to production, repairs, etc.
ANALYSIS Choice of method	Selection of methods for hazard identification and, in some cases, for probabilistic estimates.
Structuring	The object is divided up into sections at an appropriate level of detail.
Identification	Hazards are identified.
(Quantification)	Quantitative estimates of consequences and probabilities can be made in some cases.
Assessment	Risks are evaluated. Which criteria are needed?
SAFETY MEASURES Making proposals	Proposals for safety measures that will reduce the risk.
Revising proposals	The ideas are evaluated and worked through.
REPORTING Risk summary	The hazards are listed, with an account of probability of occurrence and/or consequences in some cases.
Proposals	Summary of the safety measures proposed.
DECISIONS Decision forum	Who is to draw conclusions and make decisions?
Type of decision	Implementation of recommendations, further analyses, etc.
Implementation	The decisions shall be acted upon. What are the conditions for the implementation of the changes?
Follow-up	Checking that measures as specified have been taken.

participate at important stages of the analysis, such as the formulation of aims, planning and reporting, and also when special problems arise.

A reasonable number of team members is between three and six. People with a range of skills should be included, the ideal team composition depending on what is to be studied. All members of the team do not need to be familiar with the analytical procedure.

Several types of problems can arise if just one person conducts the entire analysis. One is that it just results in a stack of paper, which is largely ignored. An important advantage in creating a team or management group lies in the way the analysis can be rooted at company level. Through a stage-by-stage process of clarification and adjustment, the results can become broadly accepted within the company. In the course of the analysis, different opinions can be expressed on the way in which risks should be assessed, on the appropriateness of safety measures, on how the system really functions, etc. The existence of a study team enables the correctness of parts of the results to be checked as the analysis progresses. It is of advantage if differences of opinion are detected early, so that they can be resolved or handled appropriately. Moreover, when people have been involved in conducting an analysis, it is more likely that they will want their conclusions and recommendations to be accepted.

Definition of aims

The objectives and subobjectives of the proposed analysis may need to be further defined. This may concern which hazards should be considered, how detailed the study should be, etc.

Setting object limits

Before starting on the analysis, a number of questions should be answered: How large a part of the object is to be covered and which system boundaries are to be specified? Which activities (e.g. maintenance, transport, etc.) are to be included? Will the analysis consider changes to equipment that have been discussed but not yet implemented?

Specification of assumptions

It is necessary to establish what can be taken for granted with respect to how the system functions. The most important assumptions of the analysis should be specified in the final report. Moreover, the correctness of these assumptions should be checked wherever possible. Some examples of assumptions (which are probably only rarely justified) are provided below:

— Drawings are accurate.

— Job instructions are followed.

— Supplier of a machine will provide an adequate solution to the safety problem.

— Machinery will be serviced at the prescribed intervals and by skilled personnel. (If quantitative estimates are to be made, this is an important assumption.)

— Personnel have the knowledge and experience to handle unforeseen situations.

Planning

It can be difficult to specify a precise schedule in advance. Among other things, it is not known how many hazards will be detected and what problems may arise. If such factors are taken into account from the beginning, changes of plan need not be regarded as failures.

Conducting and planning a safety analysis is simpler in the case of an existing installation. When an analysis is applied to planned plant or equipment, the project schedule for the installation will be the governing factor, and the safety analysis will have to be adapted to this. In addition, access to information will be poorer. Safety analysis in the context of the planning of an installation is discussed in section 13.3. A simple example of how an analysis for an existing installation might be planned is provided in table 12.9.

In some cases, it can be appropriate to prepare a written outline when planning and embarking upon an analysis. This particularly applies if the person in charge of the analysis is not employed by the company in question. For example, the outline may include the following items:

— Access to documentation.

— Tasks for different team members.

— Utilization of the results.

180

Table 12.9 *Example of the scheduling of an analysis*

Activity	Date	Responsible team member	Comments
Preparations	3/1	T	First meeting.
Aims		L	
Setting limits			
Assumptions			
Information to employees	16/1	M1	Workplace meeting, notice board.
Information			
Drawings & manuals	27/1	M2	Checking for correctness.
Interviews with 2 operators	3-7/2	L	
Reports	10/2	M1	
— Accidents			Last 3 years.
— Near-accidents			
— Disturbances to production			
Analysis			
Energy analysis	10/2	L	
Preparations for Deviation			
analysis	14/2	L	
Identification of hazards	14/2	T	
Risk assessment	20/2	T	
Safety measures			
Producing ideas	21/2	T	Designer takes part.
Reviewing ideas	24/2	T	Designer takes part.
Summary	3/3	L	List of safety measures.
Time in reserve			
One extra meeting if needed.			
		L	Seems to be problems with computer control system.
Decision	10/4	M1	Probably the head of department.
Information to employees	12/4	M1	

Key:
L = Leader
T = Team
M1/M2 = Team members

Time requirements

The time taken by an analysis can vary considerably. It will depend on the nature of the object, the aim of the analysis, the method selected, degree of familiarity with the method, etc. Estimates of the time required for some methods are provided in table 10.4.

If we restrict ourselves to the three simplest methods (Job safety analysis, Deviation analysis and Energy analysis) and assume that the team leader has a certain familiarity with the methods, the following approximations can be made. Information needs to be gathered for all three methods but the degree of detail varies. Time required for data collection can be between one hour and several days. Identification of hazards can take between half a day and a full day. These methods also involve a risk assessment stage which is implemented in one go for all identified hazards. Typically, this might take one or a couple of hours. Generating safety proposals may require a meeting of about four hours, plus some time for further consideration and revision.

An analysis of a medium-sized object using some of these basic methods will require the person in charge to devote between two days and a week of his time, possibly more. The study team will meet perhaps three times, each for half a day. But, of course, the time needed may be longer, if the level of ambition is high or if the situation and object are complicated by nature.

The identification of hazards in itself takes up only a limited amount of time, perhaps 10% – 20% of that needed for the entire analysis. Some time is taken up in completing the record sheets, following up relevant issues and providing information.

12.4 Information and analysis

The analysis requires information of different types, both on how the system functions and the problems that exist. It is perhaps simplest to obtain information by working together with people who are already familiar with the installation, e.g. through their participation in the study team. Table 12.10 provides examples of various sources of information that can be utilized.

Information on the system

Information is needed on the system and how it functions. This may be documented in the form of drawings, job instructions and maintenance schedules. However, the written documentation will describe only a limited portion of the reality. First, it will only cover certain system aspects; second, it may not be accurate. Instructions may be out-of-date, may not be followed, or can be simply incorrect.

For this reason, supplementary information is needed. This may be obtained by making observations at the installation. The knowledge of people who know the system can be accessed though interviews, or through their inclusion in the study team. Photographs and video recordings can be utilized, especially for "one-off" operations such as a repair to a piece of equipment.

Table 12.10 Examples of sources of information

Source of information
ON THE SYSTEM Drawings Job instructions Operating instructions for machines Maintenance schedules Computer programs Reliability data on components and equipment
ON PROBLEMS Reports on accidents at the installation Reports on near-accidents at the installation Reports on disturbances to production and breakdowns Information from other installations Inspection records Reports on repairs Observed damage to equipment, etc.
ON BOTH SYSTEM AND PROBLEMS Members of the study team Interviews with designers, etc. Interviews with personnel at the installation Direct observation (if the installation already exists) Experiences from other installations Photographs Video recordings

Information on problems

To include knowledge of different kinds of problems is of great value. It can be employed in a variety of ways:

— For identifying hazards.

— In the course of risk assessment. (Have similar hazardous situations arisen before?)

— For judging whether hazard identification has been sufficiently thorough.

— In discussing whether there is a justification for safety measures to be taken.

Written documentation may concern:

— Accidents.

— Near-accidents.

— Disturbances to production.

— Inspection records of different types.

— Repairs carried out.

Further information can be obtained through interviews. These can be undertaken in advance or in the course of the analysis. Photographs can be employed to provide documentary evidence that certain problems have actually arisen.

Selection of method

When knowledge of the system and an overall picture of the problems have been obtained, a more definite choice of method can be made (see chapt. 10). It is often a good idea to use more than one method. These can then be used to complement one another.

Structuring

Structuring involves both creating a simplified model of the system and its functions, and dividing the system as a whole into blocks. Several methods require the system to be structured in a specific way. This applies in particular to Job safety analysis, Deviation analysis, Energy analysis, FMEA and HAZOP. Structuring is a critical stage of the analysis and should be carried out carefully. Thought should be given to whether all important elements have been included. This can be done by making checks in relation to the boundaries of the analysis and against the assumptions made at the beginning.

Identification of hazards

The way in which hazards should be identified is linked to the method employed. It is usually best to follow the structure that has been created and identify hazards for one block at a time. Risk assessment and discussions of safety measures take place when the hazard identification stage has been completed. Two of the reasons for this are that the approach to assessment will be more consistent and that hazards that seem minor will not be dismissed at too early a stage of the analysis.

If hazard identification is carried out in a team, it is important that a creative atmosphere is generated. This means that the team leader should try to avoid criticism of proposals and assessments at this stage.

In practice, it is sometimes advantageous to relax the planned structure. Interviews and meetings generate ideas and observations that should be followed up before they are forgotten. After these sidetracks have been examined, it is still quite easy to return to the main theme established through the structuring procedure. This means that the team leader must have a grip on the main thread of the analysis and be able to give a structure to the record sheet.

Quantitative estimates

If quantitative estimates of probabilities or consequences are included, these will usually be made after the identification stage. This provides the basis for selecting the scenarios most relevant for further work.

Risk assessment

In qualitative analyses, identified hazards are assessed one at a time. It is best to make these assessments in one go, after the hazard identification stage has been fully completed. This can be done even when several methods are employed. The advantages are that the

185

assessments are more consistent and a better overall picture of the results is obtained. However, when large systems are analyzed, this approach can become unmanageable.

Practical risk assessment has been discussed in chapter 11. Before beginning, it is a good idea to establish which evaluation criteria should be employed. Should disagreement arise, it is not necessary for this to be immediately resolved. Divergent views on evaluation can be noted on the record sheet. These issues will come up later, both when safety measures are discussed and when decisions are made on which measures are to be implemented.

Discussion of safety measures may have an effect on assessment. A greater insight into a problem has been obtained, and it is also known whether the problem can be solved. But, hazards should **not** be re-evaluated and then regarded as acceptable simply because it is not possible to find a satisfactory control measure. However, the opposite might apply. If a safety measure that can reduce risk is easily available, it is reasonable that it should be implemented.

12.5 Safety measures and results

In this section, it is assumed that a list of assessed risks has already been prepared. Alternatively, a fault tree may have been constructed, and control measures are needed to address certain critical branches. A study team is of advantage, as it facilitates the generation of ideas, and the development and evaluation of safety measures.

Safety proposals

First, an attempt is made to generate ideas for safety measures. Two of the methods contain a systematic procedure for this. The more ideas, the better. Only at the next stage of the analysis is there a need to exercise greater restraint. If this can be properly explained, it is easier to obtain a creative atmosphere in the study team.

Revision of safety proposals

This stage of the analysis involves a critical examination of the proposals. They are reviewed and revised so that the most concrete formulations can be obtained. It is not possible to resolve everything at just one meeting. Some ideas require further treatment.

The following questions should be considered when developing and evaluating ideas for safety measures:

— Does the extent of the measure match the size of the risk?

— What effect will the measure have?

— Does the measure have a short or long-term effect?

— Does the measure have only a local effect? Might it only apply to a particular machine or are its effects more general?

— Are the side-effects positive or negative?

— What are the financial considerations? What will the costs be and how much income will be generated? (See sect. 13.4.)

Reporting

The content of the report generated by the study will depend on the aims of the analysis. It may consist of a risk assessment or a summary of proposed safety measures. An account of the limits set and the assumptions made should be included. The amount of detail required will vary according to the situation. For simpler analyses, an oral description and a list of recommended safety measures may be sufficient. In the case of large analyses, reporting requirements can be extensive (e.g. CEC, 1982).

Decision making

A safety analysis should finally result in the taking of decisions. The nature of the decisions may vary.

— Safety proposals are accepted (and implemented) or rejected.

— Safety proposals are to be revised, and will be decided upon later.

— A recommendation for implementation is made, which will be decided upon at a higher level.

— Further analysis of the system should take place.

187

— The analysis, in terms, for example, of its quality or the nature of risk assessment, is rejected or accepted.

Implementation

The final stage is the implementation of the decisions taken. It is of benefit if those who have taken part in the analysis can also participate at this stage. Otherwise, the considerations that lie behind an analysis can be disregarded, so that the result will not be as good as it might possibly have been.

Experience of the evaluation of safety analyses suggests that this final link is sometimes the weakest (see chapt. 14). For this reason follow-up is essential.

13 Safety management

13.1 On the components of safety management

One of the goals of companies is to obtain a high level of safety. They face demands from society that the work environment they offer is safe. Such demands are expressed in legislation, in the directives of the authorities, through the activities of the Labour Inspectorate, etc. There is an extensive body of regulations and the ones that are relevant can be difficult to find. They tend to be general by nature and cannot lay down in concrete terms what is required for a specific installation. In some cases, where plant or equipment is of a uniform type, more precise requirements can be specified.

The opportunities for an individual to prevent and avoid hazardous situations depend on his or her own work situation. Those people who have the freedom to determine the way in which they work also have a certain degree of control over the level of risk they face. But in many work situations, the opportunities for exerting influence are poor, e.g. when working on a production line. In some countries, employees have the right to refuse to carry out work which they regard as hazardous. Probably, however, it is unusual for this right to be exercised.

Scope of this chapter

It is at company level that the most important part of injury prevention work can be carried out. It is also at this level that conditions for such work are most favourable. For these reasons, this chapter is concerned with safety activities within companies.

The description that follows is general by nature, but is concerned with workplaces that can be regarded as "normal" or "common". One of the reasons for this is that most of the descriptions available in the literature refer to installations where major accidents might occur, such as within processing industries or at nuclear power plants. At such workplaces, a great deal of attention is paid to safety, both by the company in question and the authorities. At most "normal" installations, obtaining a satisfactory level of protection against accidents and other undesired events is not such a self-evident target.

There is an extensive literature on safety management and a variety of concepts, such as "risk management", "loss control" or "quality assurance", are employed. It is not the purpose of this chapter to provide a full description of the area. The reader who wishes to go into the subject in greater depth should refer to the more specialized literature (e.g. Health and Safety Executive, 1991)

Definition

Safety management in this context refers to the decision making procedure that applies in relation to accidents and other types of undesired events. This procedure includes the setting of targets, the identification of hazards and the making of decisions on which measures should be taken. The procedure can be adopted more or less consciously, and may be planned or unplanned.

The process of safety management

Figure 13.1 illustrates company management in relation to a specific target. The illustration is highly general and applies to virtually any type of activity within an organization. The target is compared with a "state" which is to be managed or controlled. This requires that the current state of an installation, e.g. with respect to accident risks, is measured. The measurements are then examined; and, if deviations are unacceptable, changes are made so that the installation obtains the desired properties. Accident hazards are often difficult to detect, which is why experience must be fed back to decision makers in a systematic manner. Safety work within a company can be regarded as an important link in the chain that provides this feedback. Safety analysis is one of several aids available for establishing this link.

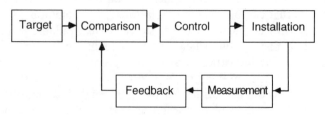

Figure 13.1 Schematic model of targeted control

We start our discussion of the general model illustrated in figure 13.1 by considering the "installation". We use the word in a wide sense to refer to the industry or procedure which is to be controlled. An installation has a number of properties, related to safety, productivity, quality, etc. If it has deficiencies, there are no automatic mechanisms whereby these are detected and corrected. "Hidden" defects must be measured in some way, and the results used within a safety management system.

Targets

Safety targets may be part of an official "company policy", although targets in this area are not always formulated. Such targets are also defined by directives of the authorities, accepted industrial practice and agreements with trade unions. Nevertheless, the drawing-up of a company policy is one way in which company management and employees can demonstrate their interest in and commitment to safety matters.

With respect to accidents, the ideal situation would be one in which none at all occurred — but this is not wholly realistic. Instead, the target might be to obtain a low accident frequency. But even this is not enough where serious physical risks are involved, e.g. when people are working at heights or use dangerous machines. Hazards need to be controlled and routines must operate in a reliable, unbroken manner. Also, changes are often made to production systems and these might lead to new risks. For this reason, the setting of targets should take place in conjunction with a consideration of the requirements for systematic safety work.

Example of a company safety policy

An outline safety policy, designed for general application, has been prepared by the Association of Swedish Chemical Industries (1985). An example of the form such a policy might take is given below.

"Safety policy

The aim of safety policy at our company is to provide people, the environment and property in and around our plants with protection from all the types of hazards which are related to our industrial operations.

Safe environmental and operating conditions are a precondition for efficient production and for the creation and preservation of goodwill and competitiveness. For this reason, we regard safety issues as having the greatest importance in the planning of plant and equipment, in routines for operation and maintenance, and also during production and service.

Company management is fully aware of its responsibilities with respect to environmental and safety matters. Together with skilled operators and supervisors, it is working to ensure the safety of both its equipment and its job routines. At the same time, every employee has a personal responsibility for his or her own safety and that of others. This responsibility is fulfilled by showing good judgement, following accepted job routines and using equipment correctly. In accordance with this, we shall see to it:

— *that all those affected, through the provision of training and information, are made well aware of potential risks and how these are best avoided.*

— *that all accidents, near-accidents and hazardous situations are reported, and that suitable control measures are implemented.*

— *that each and every one of our employees is acquainted with our emergency contingency plan and knows what he or she shall do in the event of an alarm.*

— *that we keep ourselves well informed with respect to current legislation and directives of the authorities.*

— *that all personnel, by following internal company regulations and job instructions to the best of their ability, comply with the provisions of the law.*

— *that the distribution of responsibility with respect to safety matters is clear and well documented.*

— *that the terms set by the authorities for carrying out our activities are met and that permissible emission limits are not exceeded.*

— *that all employees, the public, and local and other authorities are openly provided with constructive information on issues of safety."*

Measurement

Measurement in this context refers to the procedures and routines that can be employed to formulate and sometimes make quantitative estimates of levels of safety or other system properties. Measuring can be carried out in many different ways. It can involve compiling accident statistics, conducting safety analyses or inspections, or utilizing the appropriate channels to pass on knowledge of problems that have been experienced at plant level. An individual may know of many problems, but this knowledge is not passed on and taken account of. As a result, the problems cannot be handled in an appropriate manner.

Feedback

Results of the measurements must be formulated in a comprehensible manner. The intention is that these reach an organizational level where evaluations can be made, decisions arrived at and measures taken.

Comparison

The information obtained through the feedback of experience is evaluated and provides the basis for decisions on possible changes. If this is carried out in a rational manner, it will mean that safety targets are given a specific, concrete form. There are, of course, a number of ways in which hazards and appropriate levels of safety can be assessed at an installation. An example of a systematic approach has been presented by Rowe (1990).

Control

Decisions on changes and improvements must ultimately have an effect on the hazards to which people are exposed. This usually takes place through the company's line organization.

Small companies

The discussion in this chapter, and also in the book as a whole, applies in principle to all types of companies. But the issues raised may well have less relevance in small companies, simply because the problems they face are small in absolute terms. Nevertheless, for individuals within a small company, the risks for accident and injury can be just as great as they are in larger companies. Also, a small company is probably even more vulnerable to the effects of an employee sustaining an injury and being kept away from work as a result.

In many ways, safety management is simpler in small companies. One reason is that knowledge tends to be concentrated in the hands of just a few people, meaning that decisions are made through less formal channels. On occasions, however, it is certain that small companies would benefit from the understanding of hazards provided by a systematic approach to safety issues.

Examples of problems

The nature of safety work can vary with respect to both the extent to which it is systematic and the level of its ambition. One survey (Kjellén, 1983) of accident prevention activities at a number of companies provided a variety of examples of what could be improved. The following conclusions were drawn:

— Accident statistics provided a poor basis for taking safety measures and setting priorities.

— Accident reports were filled in in a cursory manner.

— Results were not used systematically for the prevention of accidents.

— Safety rounds tended to be devoted to the detection of technical defects. "New" hazards were seldom detected.

— If the same defect was detected on many occasions during safety rounds, nothing was done about the underlying cause.

This can be summarized by stating that the feedback of experience was considerably poorer than it should have been. The conclusion of other surveys undertaken at company level (Burström & Lindgren, 1983) is that the management of accidents risks is usually not sufficiently systematic for causal connections to be detected.

Other models

The process of safety management in general can be described in different ways. The Health and Safety Executive (1991) specifies six different main elements:

— Policy.

— Organising.

— Planning and implementation.

— Measuring performance.

— Reviewing performance.

— Auditing.

Reviewing involves the examination of performance and making decisions on how this can be improved. Auditing is the structured process of collecting *independent* information on how well safety management functions. Both of these activities provide the basis for the feedback loop needed for the entire process to be efficient. Reviewing and auditing may result in the reformulation of a policy and require changes in the other elements.

13.2 Safety work

As is illustrated by the stairway in figure 13.2, a company's level of ambition can vary considerably. The lowest level (1) refers to doing what

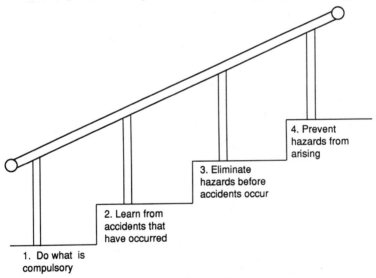

Figure 13.2 Levels of ambition in safety work.

one is compelled to do. This might even involve an attempt to avoid taking the minimum action required, so that all expenses which seem unnecessary can be avoided.

At level two, lessons are learnt from accidents that have occurred, and attempts are made to prevent them from recurring. Level 3 means that efforts are made to identify and eliminate hazards before accidents occur. Working at levels 2 and 3 involves making system changes, which means that costs are incurred.

Working at the highest level (4) means that the company attempts to ensure that hazards and other types of problems do not arise. They are taken into account when an installation is planned or about to be converted or re-built. The changes can be made on the drawing board not to already-existing machinery, which is often considerably cheaper.

Examples of accident prevention activities

A variety of different activities can be undertaken so that accidents are prevented (table 13.1). Most are related to some type of feedback, some

Table 13.1 Safety activities

1. **DO WHAT IS COMPULSORY** — Take measures after these have been pointed out by the authorities. — Take measures in the light of mandatory, unequivocal directives.
2. **LEARN FROM ACCIDENTS THAT HAVE OCCURRED** — Investigate what has happened. — Compile statistics. — Prevent identical accidents. — Prevent similar accidents. — Make systematic use of information on accidents that have occurred.
3. **ELIMINATE HAZARDS BEFORE ACCIDENTS OCCUR** — Carry out safety rounds. — Report near-accidents. — Conduct a safety analysis of existing installation.
4. **PREVENT HAZARDS FROM ARISING** — Select low-risk technical equipment. — Apply good design practice. — Conduct a safety analysis on the basis of planning documents, etc. — Ensure that the safety organization and its accompanying routines are effective.

being undertaken after an accident has occurred, others being designed to discover defects before anything has happened. Some of the approaches that can be adopted are discussed briefly below.

Accident statistics

Statistics do not in themselves prevent accidents and their compilation may easily become a formality of no practical significance. Company statistics on accidents are often designed to make comparisons between years. They sometimes show distributions by part of the body injured and age of the injured worker, but tend to provide a poor basis for prevention and the setting of priorities.

Accidents that have occurred can be regarded as a source of knowledge — which can be updated on an annual basis. The greater the number of years considered, the better is the quality of the information. It can be examined from different angles so that ideas on what action might be taken can be developed in stages. Examples of what might be included in a body of accident statistics are as follows:

— Accident frequencies. Note that these can be calculated and interpreted in different ways. Statistics can be obtained for the entire installation and are sometimes broken down by department. Comparisons can be made with the average for the industrial sector. It can be appropriate to distinguish between white-collar workers and production personnel as these categories are in different risk situations.

— Days absent from work.

— What the injured persons were doing when the accidents occurred, e.g. types of job tasks.

— Places where the accidents took place. These might be marked on a drawing of the installation.

— Deviations that preceded the accidents.

— The equipment involved in the accidents.

— Summary of the costs that have arisen as a result of the accidents.

— Summary of the measures that have been taken as a result of accident investigations.

Accident investigations

As much as possible should be learnt from accidents that have occurred. These can be investigated in many ways, and in greater or lesser depth (e.g. Ferry, 1981; Hendrick & Benner, 1987). There are several examples of good results being achieved after making improvements to investigation routines. In one case, referred to in section 15.8 the amount of information obtained on what had happened before the accidents was doubled, while five times as many safety proposals were generated. Most important of all, there was a reduction in the number of accidents that occurred.

Based on experiences from this example, the following can be recommended:

1. That a special investigation group is set up.

2. That investigations are conducted methodically (e.g. deviation investigations).

3. That relations of responsibility for safety measures are made clear.

The investigation group would be responsible for:

1. Conducting thorough investigations.

2. Analyzing the results of the investigations and proposing safety measures.

3. Referring proposed safety measures to the appropriate decision making forum.

4. Following up the implementation of safety measures.

5. Reporting results to safety organizations within the company.

Safety rounds

In principle, safety rounds and inspections provide the opportunity to obtain a comprehensive picture of hazards at a workplace. However, some of the examples referred to section in 13.1 suggest that this does not work in practice. For this reason, it might be necessary to breathe new life into safety round procedures to avoid the blindness of familiarity. Some examples of approaches that might be adopted are provided below.

1. Improvements to inadequate correction routines

We assume that there are normal procedures for taking immediate measures when technical defects affecting safety are discovered. Safety rounds should not be necessary for this. If such a defect is detected in the course of a safety round, it is relevant to ask why the defect still exists and how it can be prevented from re-occurring.

2. Other inspections

Check whether other inspection or repair routines are in operation or needed. Examine the results and investigate how the routines operate.

3. Targeted safety rounds

Safety rounds can be given a specific focus. They can be aimed at a variety of types of problems, concentrating on one at a time. The theme of a particular round might be a certain type of hazard, the functioning of safety equipment or the suitability of job instructions.

4. Task-oriented safety rounds

A section (unit operation) of the workplace might be selected for investigation in greater depth. A specific task is studied, and an attempt is made to identify particularly hazardous phases of work and generate safety measures in relation to them. The focus is on people and job routines rather than on machines. In principle, this is a Job safety analysis.

5. New employees

People just starting at a workplace may be particularly at risk. They can, for example, copy hazardous work practices that are especially dangerous for a beginner. The specific focus of the safety round is on how new personnel should be introduced to the job, but it might also give rise to improvements in how experienced workers carry out their tasks.

Near-accidents

A near-accident is an event that has almost led to the occurrence of an accident. In principle, it differs from an accident only in that an injury has not occurred. Near-accidents are far more common than accidents.

A reasonable estimate is that there might be up to 100 times as many, depending on the workplace and which measuring procedures are adopted.

The collection of data on and the analysis of near-accidents can be an excellent supplement to the investigation of accidents when measuring "safety" in a system. This is both because there are more of them and because such incidents are regarded as less controversial than when someone is actually injured. Schaaf *et al.* (1991) suggest that there are three general aims involved in collecting data on and analyzing near-accidents:

1. To gain a qualitative insight into how failures can develop into accidents.

2. To gain a quantitative insight into the occurrence of factors giving rise to incidents.

3. To maintain a certain level of alertness to danger when rates of accidents are low.

Near-accidents investigations can be conducted in the same way as accident investigations. However, special attention needs to be paid to how information on near-accidents is collected. Such information can be gathered from the "spontaneous" reports of those who have observed an incident, systematic routines or specific campaigns.

The spontaneous reports of individuals provides information on only a small proportion of the incidents that occur, usually those that are technical by nature (Kjellén, 1983). Schaaf *et al.* (1991) have discussed how systematic routines can be designed. They point out the importance of not just focusing on technical components and human behaviour, but also paying attention to organizational and managerial causes. They further stress the need to study error compensation behaviour and recovery, and the technical and organizational resources which provide support for this.

Experiences of near-accident reporting "campaigns", conducted over a short period of time (e.g. a week), have been favourable. An improved information base and a generally increased interest in safety issues are obtained. Nevertheless, experiences of the benefits of near-accident reporting are quite variable (Menckel & Carter, 1984).

Specific reasons for gathering information on near-accidents over a limited period of time include the following:

— That new plant or equipment is to be put into operation.

— That changes have been made to an installation.

— That major efforts are needed to reduce accidents in a department with a poor accident record.

When implementing a near-accident reporting campaign, the following factors need to be taken into account:

1. Careful information needs to be provided before the campaign. The reasons for the campaign should be made clear in advance. Information on results and safety measures taken should be supplied afterwards.

2. The means of obtaining information should be as simple as possible. Someone can just go round and ask, or a simple questionnaire (easy to fill in) can be used.

3. It is of benefit if human errors of different types are covered. However, this information is difficult to obtain and conscientious preparatory work is necessary. One way might be to stress the need to establish whether machines are difficult to operate or whether manuals are unclear or incomplete.

4. The reports are to be used. Setting up an investigation group may well be a good idea.

5. At the same time, the campaign can be used to obtain people's views on hazardous situations and various types of other problems.

6. The campaign may have favourable indirect effects. For example, those who participate may become more interested in safety issues.

13.3 Safety management at the planning and design stage

Technical changes designed to increase safety at an already-existing plant can be difficult to implement and may be expensive. It is of advantage to take account of safety right from the beginning. The aim of planning and design is to obtain an installation that works. In some cases, safety may be an important element, e.g. in some processing industries, but it is not usually treated as essential.

Examples of problems

A study of the planning and design procedures of three companies has been conducted (Fång & Harms-Ringdahl, 1984). These were not so effective when looked at from the perspective of accident prevention. Little interest was shown in safety issues, and these were not considered in a systematic manner. Safety was handled largely through inspections, during and after installation.

Schedules were tight, meaning that little time was available for the analysis of hazards and possible control measures, even where there was a will to implement such measures. On the other hand, the projects had histories which went back several years. Investment proposals tended to be made at one point in time, but several years elapsed before they were put before the companies' boards of directors. This suggests that the time available could be better utilized.

Responsibility for safety issues was diffuse. With respect to one and the same project, there were different views on who should ensure that the risks for accidents were not too great. There was an unclear distribution of responsibilities between project supervisors, suppliers, safety engineers, etc. Follow-up activities at the completed installations were not undertaken systematically.

Planning and design procedure

The figure illustrates a model of how planning and design might progress. The process starts with a planned target, encompassing functional requirements, financial considerations, project schedule, etc. But it may also include a company policy which needs to be given a concrete form for the project in question.

On the basis of the target, proposals are made for its achievement. These are evaluated, and implemented if regarded as satisfactory. Otherwise, they must be revised. After implementation, the final result is evaluated. When this is approved, the process of planning and design has been completed.

The model provides just an outline of the planning and design process. In reality, it is much more complicated, involving a large number of different activities, which interact with one another. But one main thread, that of proposal — evaluation — implementation — evaluation, runs through many of the subactivities.

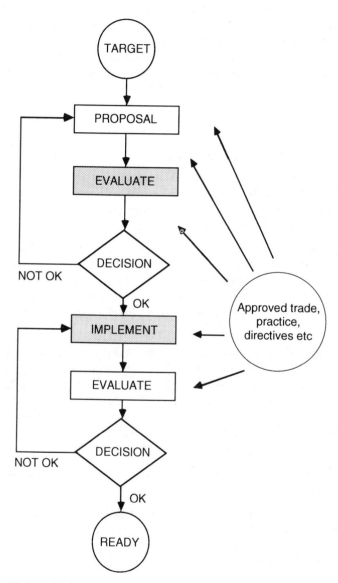

Figure 13.3 A model of planning and design procedure

The basic idea, of course, is that an attempt should be made to find a solution that is as good and safe as possible right from the beginning. Such solutions are based on experience, different types of norms, good design practice, etc. Safety issues are sometimes considered, but only to a limited extent. For some industrial sectors, lists of what should be considered in the course of planning and design have been compiled, e.g. for the chemicals processing industry.

Computer-controlled equipment

An ever larger proportion of industrial equipment is becoming computer controlled. As already discussed in section 2.3, safety issues at such installations are of importance. Some points with special relevance to planning and design are repeated here.

Several studies (Backström & Harms-Ringdahl, 1984; Döös & Backström, 1990) have shown the handling of disturbances to production to be the most hazardous type of job task. At the same time, it is common for companies to be poorly prepared for taking corrective action of this sort. Job instructions tend not to be available, which permits more or less hazardous practices to develop.

In simple terms, the explanation may be that it is assumed at the design stage that automated equipment will always function as planned. As a result, the task of the operator is regarded as being so simple that careful preparation for it is not required. But there are a large number of measures, many of which are relatively simple, that can be taken at the planning and design stage (e.g. Backström & Harms-Ringdahl, 1986). Some examples of what might be considered are provided below:

1. Disturbances will occur. Planning is needed for corrective procedures to be safe and rational.

2. To identify deviations, which might lead to disturbances and hazards, simpler forms of analysis can be conducted as soon as the principles for automatic control are ready. For example, Deviation analysis can be employed even at this stage.

3. There is a need for manuals which are tailored to the needs of the operator. These should contain descriptions of both normal work and what to do in the case of disturbance.

4. It must be easy to stop the hazardous movement of machine parts, and then to re-start the machinery.

Operational readiness

For an installation to be regarded as ready for use, a large number of different phases and parts must be fully operational. Usually, satisfactory attention is paid to technical aspects of the system, but other aspects tend to be neglected. In addition to the technical components, there must be people with adequate training, a variety of management, supervision and maintenance procedures, etc.

There is a methodology available, called "Operational Readiness" (Nertney *et al.*, 1975), which stresses the need for an overall picture of the final result. From the beginning it was developed for relatively advanced systems. It is not really a method for safety analysis, but rather a planning aid.

For an installation to be ready, all its component parts must also be in operating order. The methodology prescribes the use of check lists to ensure that as little as possible will have been neglected when the plant or equipment is put into operation.

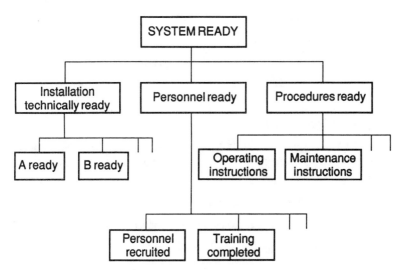

Figure 13.4 *A simplified version of Nertney* et al.*'s control tree for "operational readiness"*

What to consider at the planning and design stage

Work for the reduction of accidents that is carried out at the planning and design stage must be systematic. Examples of the actions needed are listed below. Some items on the list might seem obvious. Nevertheless, they are easily overlooked:

1. Define a safety target.

2. Decide the level of ambition. Might serious hazards arise?

3. Clarify issues of responsibility.

4. Make a list of the norms and directives of the authorities that apply to the planned installation.

5. Allow time for the work. If company management plans in advance, decisions do not need to be made in a rush.

6. Take account of safety throughout the planning and design period.

7. Work systematically, e.g. by using either general or tailor-made check lists. It may be possible to conduct safety analyses.

8. Try to adopt a systems perspective when approaching safety issues.

9. Follow up problems and incidents that arise when the installation is first in operation.

10. Be prepared to remedy possible defects, e.g. by reserving funds for the period that immediately follows system start.

The role of safety analysis

Safety analysis is used to study proposals that might give rise to hazards or defects. The situations where such analysis might be most appropriate are:

1. When the project is being planned.

2. When safety targets are being given concrete form.

3. During the examination of proposals.

4. At final inspection.

5. During the initial period of operation.

In addition to the general features of planning and design discussed in chapter 12, there are further things to be considered if maximal use is to be made of safety analyses.

SCHEDULING

1. Identify the occasions on which decisions will be made and when the results of analyses will be needed, such as before equipment is purchased or before the production layout is settled.

2. Check how analyses can be fitted into the schedule. Will there be time to utilize proposals for improvements? Reports of analyses must be made available in good time so that they can affect the decisions taken.

INFORMATION

3. Only scanty information may be available at the beginning, but preliminary analyses can still be conducted.

4. Preliminary proposals can provide the basis for a rough analysis. The opportunities for making improvements are greater at an early stage, as less work has been put into the project.

PURCHASING

5. When machines are ordered from a supplier, it is often the case that only functional requirements are specified. This means that the machine must meet the requirement specification, but might have any one of a variety of detailed designs. A specific clause can be inserted in the purchasing contract to ensure that the results of safety analyses and proposals for change can be incorporated at a later stage.

AN OVERALL PERSPECTIVE

6. A completed installation is not just a piece of technical equipment. What people will do and what procedures will be adopted are also important, for safety as well as for production.

7. The analyses can consider other system features as well as accidents. These may be ergonomic (in either a narrow or broad sense), or related to chemical health hazards, disturbances to production, etc.

13.4 Economic aspects

Humanitarian and ethical values play a major role in discussions of the prevention of occupational injuries. In practice, however, great weight is placed on financial considerations. Above all, safety measures might be regarded as far too expensive to implement. This particularly applies when attention is paid to costs alone, which are relatively easy to estimate. Benefits and potential financial gains are often rather more intangible and tend to be neglected.

This section considers financial aspects of accidents, safety measures and the utilization of safety analyses. It describes a simple method for estimating costs and benefits (Harms-Ringdahl, 1987b, 1990) and a brief account of the financial calculations made in a number of case studies. The studies are described in greater detail in chapter 15.

Calculation of costs and benefits

Table 13.2 provides examples of different types of costs and benefits. The table might be used, for example, as a check list after an analysis has been conducted and proposals for safety measures are available. It is then possible to demonstrate if it is profitable to implement a program of measures. Alternatively, a cost-benefit analysis might be conducted after measures have been implemented — to establish, for example, whether it is profitable to devote resources to safety analysis. In some cases it is relatively easy to couch estimates in monetary terms, in others it is more difficult.

The principle underlying such economic appraisals is that an *original system is compared with the changed system* that results from applying the results of the analysis. The "original" system may be an existing installation or a design proposal.

One way of applying the check list is to go through the items on the list and assess which are relevant to the case under study. An attempt is then made to estimate their monetary values.

Investment in the safety analysis

The costs of the analysis itself consist primarily of the working time it takes, plus certain extra costs for obtaining/producing information. The benefits lie principally in improved knowledge and a better basis for

Table 13.2 Costs and benefits of a safety analysis at company level

Activity	Comments
INVESTMENTS	
The safety analysis	
– Costs	
Time	The major part of the cost of an analysis is the time it takes.
Other costs	For documentation, travelling, etc.
+ Benefits	
Knowledge	A better overall picture of hazards and a basis for more rational decision making.
Training	Those who take part in the analysis obtain knowledge that can be applied in other contexts.
Imputed benefits	Even without an analysis, resources would still have had to be devoted to the problem.
Investments in the installation	
– Costs	
Equipment	The cost involved in implementing the technical part of the proposal.
Planning, etc.	The cost of the organizational part of the proposal, e.g. producing manuals and training expenses.
Production losses	Production lost while changes are implemented or losses from delays in introducing a new system.
+ Benefits	
Cheaper technical solutions	Cheaper solutions than those specified in the original plans.
Shorter start-up period	Analysis at the planning stage may shorten run-in times, e.g. through the anticipation of problems.
More rational planning	Fewer changes need to be made to the installation after it is ready, e.g. to meet safety requirements.
OPERATIONS	
– Costs	
Reduced production	Slower production runs as a result of the introduction of safety measures.
Other costs	For example: increased maintenance costs, heavier energy consumption.
+ Benefits	
Fewer accidents	The monetary value of a reduction in accidents largely depends on the calculation model.
Improved working conditions	For example: improved job performance, reduced risk for human error, less absenteeism.
Improved production efficiency	The production flow can be made more efficient, disturbances can be handled more effectively, etc.
Reduced risk for disturbances	Reduction in the likelihood of breakdowns and production standstills.

decision making. Usually, it is difficult to attribute monetary values to these benefits. When some of the measures generated by the analysis are implemented, the net benefit in monetary terms is in principle the difference between the income generated by and costs incurred in implementing the proposal.

A further benefit of a safety analysis is that it can raise the skills of members of the team that carries out the study. This can be compared with what it would cost to obtain the same improvement by other means. Another conceivable effect is that it improves the expertise of designers and other personnel, whose ability to anticipate and prevent various kinds of problems is increased. Moreover, there may be savings, in that the analysis acts as a substitute for investigations that would otherwise have been necessary.

Investments in the installation

Making an investment involves costs. These may be for the purchase of equipment, or in time spent on design, etc. Administrative costs may be incurred for the production of manuals, job instructions, etc. Furthermore, making changes may give rise to production losses: plant may have to be at a standstill for a time, or the implementation of an investment project might be delayed.

But an analysis can also provide benefits in the form of savings — especially when it is conducted at the planning and design stage. Technical facilities and production methods that are cheaper and simpler than those originally envisaged might be discovered. Run-in times can be reduced if start-up problems are anticipated. When hazards are discovered at the drawing board, changes are easier to make. Planning will be more rational if risks are treated at the same time as technical solutions.

Operations

Making changes to a system can have unfavourable economic effects for different reasons. It may be that the rate of production is reduced, or that a speed limit is imposed on transportation vehicles. Increased maintenance requirements may also involve greater costs. But table 13.2 also contains four types of benefits that might result from system changes made after a safety analysis has been conducted.

A reduction in the number of accidents has an economic value. The costs of accidents are discussed in section 1.3, which also provides some examples of the types of costs which should be taken into account in the calculation.

Working conditions may be improved and task-related problems solved, which can lead to improved job performance and higher work quality. It can also give rise to a reduction in absenteeism, both through reducing the risks for specific occupational diseases and through a general improvement in well-being.

The improved production efficiency resulting from a system which has been well thought through means higher productivity.

Risks for production disturbances can be identified. If their occurrence can be reduced or ways of handling them improved, financial gains will accrue.

A reduced probability for breakdowns can be allocated a monetary value. For example, it may be possible to assume that the frequency of failure is reduced from one in five years to one in ten. If the costs of breakdown can be estimated, it is simple to calculate the resulting monetary value. Such calculations are always uncertain, but can still be used to provide estimates.

Externalities

Fewer accidents also have a so-called "external" value, i.e. a value that cannot easily be measured in monetary terms. Take the example where a company makes an effort to live up to its safety policy. Demonstrating to employees that the company will take action to deal with occupational hazards improves labour relations and promotes greater job commitment. For there to be such effects, it is necessary that something is really done and that the company's striving to achieve results is made known to its employees.

Appraising the costs and benefits of an analysis

Making a cost-benefit appraisal involves many assumptions and some guesswork. In some cases, statistics are available on standstill periods, quality problems, accidents, etc. These can be employed to make cost estimates.

Different people may come to very different conclusions. But even if there is a great deal of uncertainty, such appraisals are of value when used judiciously. One way of handling the uncertainty is to present results in the form of two monetary values — representing the upper and lower limits of the economic effects of a safety analysis.

Remember that the appraisal will always involve making a comparison between the system as it was before the safety analysis was conducted and the changed system that results from the analysis.

Economic appraisal

Investment appraisal involves comparing the immediate expenses incurred in making the investment with prospective earnings that will accrue over a number of years. Some variables have to be defined in order to make the comparison:

a average annual earnings

n life of investment, number of years

r discount rate

The earnings that will be obtained from the investment over its entire life can be used to calculated its current capital value (C). Inflation effects have not been taken into account.

$$C = a \, ((1 + r)^n - 1) / (r \, (1 + r)^n)$$

Table 13.3 *Factors for conversion of annual earnings to current capital values at different discount rates and for different discounting periods*

Discount rate	3 years	10 years
5%	2.7	7.7
10%	2.5	6.1
15%	2.3	5.0
20%	2.1	4.2
25%	2.0	3.6

Some examples

Cost-benefit appraisals have been conducted for five of the examples of safety analysis presented in chapter 15. They demonstrate different ways in which appraisals can be made and take a variety of considerations into account. Depending on which assumptions are made, different results can be obtained. As the examples show, these can vary considerably. For this reason, two possible appraisals are shown for each example. Alternative I represents a cautious estimate, while Alternative II is rather more speculative, with respect to both earnings and expenses.

The results of the five appraisals are summarized in table 13.4. All figures have been rounded so as not to give any false impression of ac-

Table 13.4 Five examples of the estimated costs (–) and benefits (+) of safety analysis (values at the time of investment in SEK 000 at 1991 prices)

Example	Alternative	Net investment cost (inc.safety analysis)	Earnings (discounted value)	Profit	Net cost of analysis
A. Planning at a paper mill (sect 15.2)	I	– 700	+ 1 900	+ 1 200	– 75
	II	+ 2 200	+ 2 400	+ 4 600	– 75
B. Purchase of automatic packaging equipment (sect 15.3)	I	0	0	0	0
	II	+ 200	+ 900	+1 100	+ 30
C. Computer-controlled materials handling system (sect. 15.4)	I	– 100	+ 600	+ 500	– 40
	II	– 400	+ 1 200	+ 800	– 40
D. Section for the mixing of chemicals (sect. 15.5)	I	– 200	+ 2 200	+ 2 000	– 20
	II	– 200	+ 8 300	+ 8 100	– 20
E. Accident investigation routine (sect. 15.8)	I	– 50	+ 100	+ 50	– 20
	II	– 100	+ 550	+ 450	– 20

curacy. The amounts are denominated in Swedish kronor (SEK). At the end of 1991, 1 US dollar was equivalent to around 6 SEK. The value of the future flow of earnings and expenses at the time of the investment has been calculated using a discount rate of 10%. The investment period has been assumed to be 10 years except in case E where it is 3 years (see sect.15.8).

Some plus signs appear in the net costs columns (for investments and conducting the safety analysis). Here it should be remembered that the calculation reflects the difference between conducting and not conducting a safety analysis. For example, in case A the net investment cost figure is + 2 200. This reflects the fact that the cost of constructing the new installation was reduced by decisions resulting from the safety analysis. The differences between estimates I and II are generally rather large and reflect the different assumptions employed in the calculations. This is explained in greater detail in chapter 15.

Profitability

Conducting a safety analysis had a favourable effect on company finances in all five cases, but this does not reflect a deliberate decision to only present analyses which had positive economic results. The author has experience of a number of other appraisals, all of which have been economically favourable for the company concerned.

In all the cases exemplified, the primary purpose of the analysis was to analyze and reduce accident risks. However, it is only in the last example that major economic gains accrued from reducing the number of accidents. In this case, the entire economic benefit of the analysis lay in accident reduction. In the four other cases, the major economic benefits are related to production. This provides a strong argument for adopting an overall approach to safety analysis, i.e. an integrated approach that encompasses both the work environment and production. It means that good proposals can be backed by powerful financial arguments.

13.5 Integrated approaches

Company managements aim to attain a variety of different goals. As well as maintaining profitability through the quantity and quality of their production, they must meet the requirements of the authorities and satisfy the wishes of their employees. Thus, they must be able to cope with a broad spectrum of problems. It is difficult for an individual person to obtain an overall picture, which means that safety management and problem-solving may be only weakly coordinated.

However, improved coordination can have a number of advantages. Problems with a variety of consequences may have common causes. Examples of different types of consequences include:

— Occupational accidents and diseases, high absenteeism.

— Fires and explosions.

— Production shortfalls.

— Damage to equipment.

— Quality problems.

— Environmental damage.

Examples of causes include:

— Technical failures, especially when combined with inadequate maintenance routines.

— Various types of human error.

— Lack of knowledge and motivation among personnel.

— Inadequate correction of detected problems.

— Inadequate specification of requirements at the planning and design stage.

One conclusion is that an integrated approach to safety management might offer an opportunity to come to terms with problems in a way that is both more effective and economically efficient. For example, the ISO 9000 standards for quality (ISO, 1987) and for the management of the work environment according to the principle of internal control are comparable in many respects. They employ the same key words and phrases — such as responsibility, policy, control system, documentation, follow-up, etc.

Safety management also involves consideration of the prevention or amelioration of consequences and the planning of emergency operations. This can be seen as part of an overall picture.

An integrated approach to safety management may also have clear financial advantages. Most of the examples given in chapter 15 demonstrate that economic gains from safety analysis do not lie in improved safety but in improvements to production. This shows the importance of adopting an overall perspective, even when the only objective is to reduce risks for occupational accidents.

Choice of methods

When conducting a safety analysis, taking an integrated approach does not require the adoption of any peculiar type of method. For example, the author's experience is that Deviation analysis can be employed even when a broad perspective is adopted. Instead of searching specifically for deviations which might lead to injury, those that give rise to disturbances to production and problems of product quality can also be considered. At the evaluation stage, it is possible to make parallel judgements as to whether any particular deviation might lead to injuries, production or quality problems, or environmental damage. An example of an integrated analysis is described in section 15.7. Production and quality problems can usually be evaluated in financial terms.

In comparison with more common ways of examining production systems, safety analysis places extra emphasis on how the human being will act within a system and takes account of a variety of organizational aspects of the system. Systems are often optimized in a technical sense, but other features tend to be neglected.

14 Sources of error, problems and quality in safety analysis

14.1 Introduction

The aim of this chapter is to discuss the quality of analyses and various factors that might lead to a deterioration in results. There is no intention to provide a full account of quality issues in relation to safety analyses. The interested reader is referred to the more specialized literature (e.g. Suokas, 1985; Suokas & Kakko, 1989; Rouhiainen, 1990; Suokas & Rouhiainen, 1992). This chapter can be used as a basis for examining completed safety analyses. It may also provide a basis for anticipating and preventing problems when such analyses are planned.

Safety analyses have been subjected to criticism with respect to matters of quality, interpretation and use. The bulk of the publications that take up issues of defective analysis focus on applications to major hazards. For example, the criticism may be directed at probabilistic estimates which are based on unreliable data for the frequency of component failures and human errors, at a lack of completeness in hazard identification (Suokas, 1985), or at problems involved in the estimation of consequences (e.g. Britter, 1991).

Many features of a safety analysis can be considered in the course of evaluation. In general, quality is expressed in the appropriateness of an analysis for the application in question. As there are a large number of different applications, it is not possible to find any universal measures of quality.

The account in this chapter is based on the assumption that the aim of an analysis is to obtain a safe system, i.e. the implementation of safety measures is included. Various weaknesses in implementation are considered from two somewhat different perspectives. The first starts with the procedure of analysis and considers which deficiencies might arise. The second is based on the assumption that an accident has occurred, the investigation being into why the analysis has not prevented it.

The procedural approach

One possible starting-point for an evaluation is the safety analysis procedure itself. The various stages involved and the deficiencies that might arise are examined from a procedural perspective (Harms-Ringdahl,

1987*a*; Rouhiainen, 1990). A recent proposal for a Norwegian standard for safety analyses (SINTEF, 1991) is primarily based on this approach. The general standard for quality assurance (ISO, 1987) is also based on the idea that a suitable procedure is followed and documented. Section 14.2 contains a simple example of this form of evaluation.

Other approaches

It is possible to conceive of a number of other bases on which a safety analysis can be evaluated (e.g. Suokas, 1985; Rouhiainen, 1990). The evaluation may focus on the accuracy with which hazards are identified, as a function of the method adopted or the skills of the analyst. It may also concern the precision with which probabilities or consequences are estimated.

Such an evaluation can be conducted by comparing different analyses of the same object, carried out by different teams and/or using different methods. Another procedure is to compare the results of an analysis with its actual outcome, in terms, for example, of the accidents and near-accidents that occur. Yet another approach involves examining the theoretical basis of the analysis, e.g. the way in which estimates of consequences or probabilities are made.

14.2 Examining the analytical procedure

In the course of planning an analysis, an attempt can be made to identify problems in advance. This section provides an example of how a check list can be used to carry out such an examination (Harms-Ringdahl, 1987*a*; 1987*b*). A list of this type can also be used to evaluate an analysis after it has been conducted.

The point of departure is the analytical procedure (fig. 3.1 and table 12.8). Problems and deviations that may arise are noted for each stage, so such an evaluation is in principle a form of Deviation analysis. However, as a safety analysis is not a production system, the classification of deviations commonly used in Deviation analysis is not directly applicable.

Examples of different types of problems are listed in table 14.1. The list is not comprehensive, but is designed to indicate the types of issues that may be relevant, depending on the situation in question. Far more detailed lists can be prepared for the evaluation of specific analytical situations (e.g. Rouhiainen, 1990).

218

Table 14.1a *Examples of deviations that may arise in a safety analysis*

Activity	Deviation	Comments
PLANNING Aim & object limits	Too narrow	Hazards or subsystems not included.
	Too broad	Cannot cope with the analysis, superficial.
	Unclear	See above. The work can be difficult to implement.
Planning	Too little time	Tight schedule, incomplete analysis.
	Not enough help	Person who should help not available.
Information program Negative attitudes.	Poorly rooted at the workplace	Active or passive resistance.
GATHERING INFORMATION On the system	Information unavailable	Certain parts of the system cannot be analyzed.
	Incomplete information	Risk for serious mistakes.
	Faulty information	May apply to drawings, instructions, etc.
	Fault in the model	Assumption that the system functions in a certain way.
	Change in the system	Analysis does not apply, as major changes have been made.
On the problems	Information lacking, incomplete	Under/over-estimation of risk. Low motivation for risk prevention. Not an aid to hazard identification.
ANALYSIS Selection of method	Inappropriate	Neglects certain aspects, e.g. the role of the individual, or organizational activities. Too time-consuming, requires excessive resources. Too superficial.
Structuring	Object too narrowly defined	Incomplete analysis.
	Parts or activities excluded	Incomplete analysis.
	Division into sections too crude	Incomplete analysis, superficial.
	Too detailed	Too complex, takes too much time.
Hazard identification	Lack of skills	Inexperienced analysts, poor creativity.
	Sections neglected	Random errors.
	Stressful situation	Lack of time, poor team atmosphere, etc.
	Deliberate omissions	Hazards dismissed as unimportant at far too early a stage.
Risk assessment	Overestimation	Serious risks get lost in a mass of minor risks
	Underestimation	Important risks disappear.
	Disagreement within the team	Takes too long, etc.

Table 14.1b *Examples of deviations that may arise in a safety analysis*

Activity	Deviation	Comments
SAFETY MEASURES Proposals	Lacking Inadequate	Insufficient basis for proposals provided by the analysis. Skills lacking in the team. E.g. ineffectual measures proposed.
Revising proposals	Good proposals excluded Poor recommendations	Too expensive, involve too much change, etc. Take resources from better measures May give an impression of triviality.
DECISIONS Documentation, presentation	Incomplete, deficient	Poor impression given, proposal rejected.
Proposals	Good proposal rejected Poor proposal accepted Decision unclear Wrong decision forum Decision at wrong time	Hazard remains. Takes resources from better measures. Uncertainty in implementation. May not be possible to implement decision. Critical at the planning and design stage. May not be time for it to be implemented.
IMPLEMENTATION At executive level	Not carried through — Lack of resources — Unclear responsibility — In breach of decision Delayed	Inadequate follow-up. Skilled personnel lacking. Inadequate financial resources. Competition from other activities within the company. Decision unclear. A different opinion may be held by the executive responsible.
At workplace level	Implementation failure	Inadequate description of measures. Inadequate skills. System changed, but measures not adapted
ANALYSTS, ETC. Planning	Not carried out according to plan Delay in implementation Failure to obtain required resources Team members fail to come to meetings, etc.	Specially important at the planning and design stage. Information lacking, difficult to reach decisions.
Identification, etc.	Inadequate quality Proposals do not gain a hearing, etc. Study team does not function	Inexperience, insufficient skills. Inadequate support, not rooted at workplace level. Inappropriate composition, lack of motivation.

14.3 Accidents — despite safety analysis

Even when a system has been analyzed and measures taken, accidents may still occur. A safety analysis can be evaluated in the light of an accident that has occurred, or on the basis of an imaginary accident. Figure 14.1 shows a tree of different conceivable deficiencies (Harms-

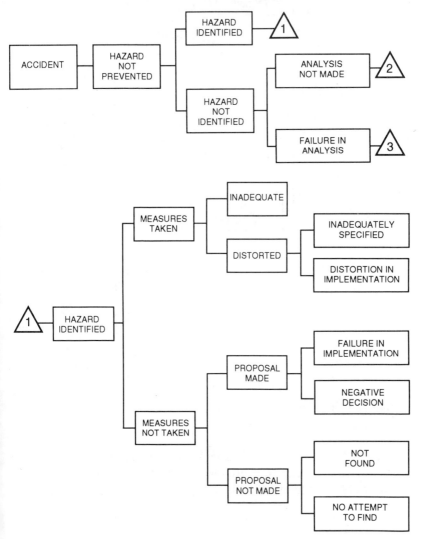

Figure 14.1 Tree for causes of accidents despite safety analysis

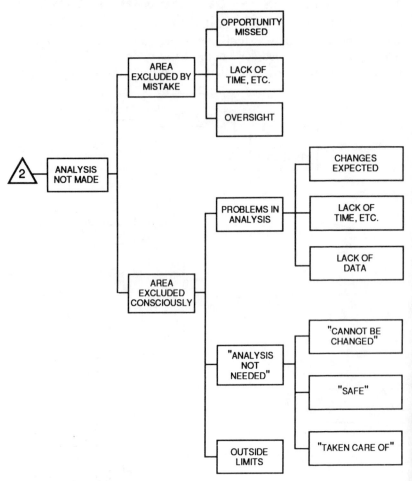

Figure 14.1 *(Continued)*

Ringdahl, 1987*a*). In fact, it is a fault tree consisting only of OR gates (which have not been explicitly marked in the tree). A number of the problems referred to in table 14.1 reappear in the tree.

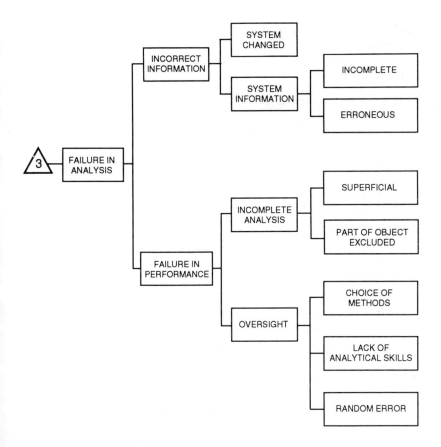

Figure 14.1 (Continued)

14.4 Some experiences

A safety analysis (described in sect. 15.2) was conducted and followed up later. Over a two-year period, 7 accidents and 8 near-accidents were recorded. Explanations for why these occurred (from a safety analysis perspective) are categorized in table 14.2 (Harms-Ringdahl, 1987a).

The table represents a simplification of the tree shown in figure 14.1. Some of the accidents can be accounted for by several different explanations simultaneously. The selection in the table is based on the principle of starting from the top (i.e. with hazards that were not identified).

Table 14.2 *An example of the relationship between accidents/near-accidents and a safety analysis*

Explanation	Number	
HAZARD NOT IDENTIFIED		9
Subsystem not analyzed	5	
Subsystem changed	4	
Failure to identify	0	
HAZARD IDENTIFIED		6
Proposal distorted	6	
Proposal rejected	0	
Measure not proposed	0	
TOTAL	15	15

One of the problems shown by the example is that proposals for changes are not always adequately implemented. Other problems include subsystems not being analyzed and changes being made to subsystems after the safety analysis has been conducted. It should be noted that none of the cases has been categorized as involving an oversight at the identification stage, or as a failure to produce some kind of countermeasure. This does not necessarily mean that these stages of the analysis were wholly successful. There may be problems which have still not manifested themselves in the occurrence of an accident.

A similar test has been applied to analyses of machines in the paper industry. Identified hazards were related to accidents that had occurred when using a large number of different machines (Suokas, 1985). 78% of these accidents had been covered at the identification stage of the analyses.

Dependence on the analyst

The analyst (team leader) is obviously a key person who influences the quality of all stages of the analysis. As discussed in chapter 12, the skills and attitudes of the study team are also important.

Hazard identification is an activity which is sensitive to the skills of the analyst and the team. The problems that arise in the course of identifying hazards have been examined in several studies. In one case (Suokas,

1985), three analyses of the same object were compared. Only 26% of the identified hazards had been included in all three analyses.

Dependence on methods

Different methods focus on different hazards and problems. Which aspects are covered by which methods are described in the summary of methods presented in section 10.1. From this, it is possible to draw conclusions on aspects that may be neglected. Selection of method is important for just this reason. Comparisons between methods have been made, among others by Suokas (1985) and Taylor (1982). It is usually of benefit to employ supplementary methods.

In the case of occupational accidents, one approach is to start with Energy analysis and then continue with Deviation or Job safety analysis. On some training courses in safety analysis, the results of adopting this approach have been checked against a dozen or so such analyses. Only about one-third of the identified hazards were covered by both methods (the shaded area in fig. 14.2). Another way of expressing this is that the employment of a supplementary method will enable roughly 50% more hazards to be identified.

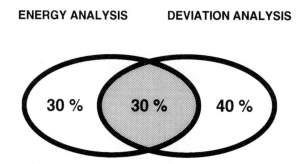

ENERGY ANALYSIS   **DEVIATION ANALYSIS**

30 %   30 %   40 %

Figure 14.2 *Examples of coverage of hazard identification using Energy and Deviation analysis*

Quantification problems

The problem of uncertainty is often great when quantitative methods are employed. For example, for a number of analyses applied in the chemicals processing industry, a comparison has been made between failures that have occurred and those that were anticipated (Taylor, 1981). The ratio varied from 0·4 to 1 680.

There is an interesting comparison of the different ways in which analyses of the same object can be conducted (Contini *et al.*, 1991; Amendola *et al.*, 1992). In a benchmark study, an ammonia plant was analyzed by 11 different teams, each with the task of conducting a complete risk analysis — from hazard identification to the evaluation of individual risk contours. The results obtained varied considerably. The principal explanations for the differences lay in the following factors:

— Different approaches to risk analysis and its implementation.

— Major variations in reliability data. For certain failures with serious consequences, component reliability data varied by several orders of magnitude.

— Discrepancies in the assessment of human success probability.

— Differences in source term definitions, i.e. with respect to assumptions on how the emission of ammonia takes place.

— Major variations in dispersion calculations.

15 Examples of safety analysis

15.1 Introduction

This chapter describes six simple examples of safety analyses and two trials aimed at improving safety work within a company. The aim of the chapter is to show that analyses can frequently be conducted using relatively simple methods and involve only a small amount of work. The examples have been selected to illustrate choice of method, analytical design, time spent on the analysis and the results that can be obtained.

The examples concern:

A. Planning and design for the rebuilding of a section of a paper mill (sect. 15.2).

B. Purchase of automatic packaging facilities (sect 15.3).

C. Examination of a computer-controlled system for materials handling (sect 15.4).

D. Examination of a section for the mixing of chemicals (sect. 15.5).

E. Control of a line of sheet-metal presses (sect. 15.6).

F. An integrated analysis of a chemicals processing plant.

Two further examples demonstrate attempts to develop safety routines within a company:

G. Investigation of accidents (sect. 15.8).

H. Improvement of safety rounds (sect. 15.9)

The bulk of the examples come from the author's own applications of safety analysis, but in cases D, E and H the work was carried out by safety engineers at the company in question.

Economic appraisals

Economic appraisals were undertaken in five of the cases. These were prepared in accordance with the principles described in section 13.4. Alternative financial estimates were made in each case. The first is the more cautious of the two, and relatively low values are assigned to costs and benefits. The second assigns higher values to some items, based on estimates that are rather more theoretical. This illustrates the uncertainties and difficulties involved in making appraisals of this sort.

In cases A, B, C and G, the estimates of costs and benefits were made by people at the company, in conjunction with the author. In case D, the appraisal was carried out by the safety engineer, who based his estimates on information made available by the company.

The estimates do not take externalities into account and are based solely on income accruing to and costs incurred by the company. In Sweden, the injury compensation costs of the employer are wholly covered by an insurance policy or equivalent scheme. Moreover, the insurance premium is independent of the number of accidents that occur. For this reason the costs of accidents would have been higher if those borne by individuals and society had been included.

Calculations were made during several different years, so all amounts have been indexed at 1991 prices. Costs have been denominated in Swedish kronor (I US dollar = approx. 6 Swedish kronor). In order to give concrete form to the relatively abstract information provided in the tables in section 13.4, the economic appraisals are discussed in a fair amount of detail.

15.2 An analysis at the planning and design stage

There had been a high frequency of accidents in one production section of a paper mill over a period of many years. A decision had been made to rebuild the section, partly because of the high accident risks. In conjunction with the rebuilding, safety analyses were conducted of various parts of the section (Harms-Ringdahl, 1982; 1987a). A description of the analysis as a whole and a summary of its economic effects are provided here. Examples of different parts of the analysis are presented below.

Analytical procedure

The rebuilding was monitored right from the pre-planning stage to the completion of the installation, a period of about a year and a half.

There was a working group concerned with the work environment and safety. The group scrutinized the results of the analyses and produced recommendations for control measures. These were then forwarded to the project management group, which accepted the recommendations in most cases. The remodelling of one of the machines, however, was postponed for several years.

Examples of the subsystems analyzed include:

— Layout and materials transportation.

— Paper machine (to be purchased).

— Paper machine (to be remodelled).

Careful planning was required so that the analyses could be carried out at the right time. This meant getting underway only when proposals of sufficient detail were available but also in good time before decisions were due to be taken. On the whole, this was managed successfully. Problems arose when some parts of the materials transportation system were to be purchased. There was not enough time to conduct the analyses, and subsystems were bought without having being examined from a safety perspective.

Changes in the frequency of accidents

The rebuilding of the section led to a reduction in the number of accidents. Information on accidents is shown in figure 15.1. The rebuilding took place during the first half of 1982.

A comparison has been made between the four years preceding the rebuilding of the section and the five years that followed (excluding the year when rebuilding took place). The average number of accidents fell by 55%, which is statistically significant ($p < 0.001$). Absence from work as a result of accidents fell by 40%. In other sections of the paper mill, the number of accidents fell by 20%, while absence rose by 10%.

Some of the accidents that occurred after rebuilding involved the machine that had not been remodelled. A further hazardous phase of work, for which a truly adequate solution was not found, was the handling of rolls of paper. Some of these accidents have been investigated.

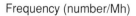

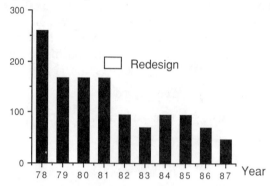

Figure 15.1 *Accidents before and after rebuilding (frequencies in accidents per million working hours)*

The reasons why they occurred despite safety analysis are described in table 14.2.

Economic appraisal

A summary of the economic analysis is presented in table 15.1. This takes up the overall costs and benefits of the safety analysis. Discussions of the various financial items that are specifically related to the different subsystems follow below. (See also pages 234 and 236.)

The analyst devoted about six weeks of his time to work directly related to the project, while meetings and other forms of participation in the analysis took up about 8 man weeks of the time of personnel at the mill. It was estimated that roughly the same amount of time would have been expended at meetings, etc., even if design and planning had been carried out in the usual manner (i.e. without a safety analysis). Some of those involved maintained that the analysis actually saved time, as the work was conducted with greater efficiency. Thus, it can be said that the safety analysis did not involve extra work for employees.

It was considered that working with safety analysis had an educational effect, corresponding to training of one week per team member, i.e. a total of 6 man weeks. Thus, the net cost of the analysis is the difference between the cost of the time spent by the analyst and the value of the training. This was set at 75 000 kronor.

The analyses did not give rise to any delays in the project, which was implemented on schedule. For this reason, no such costs have been included in the calculation.

The average number of days absent as a result of accidents was 91 per year before the rebuilding of the workshop. This fell to 56 after conversion, meaning 35 fewer days of absence per year. At this paper mill, the cost of replacing an employee was estimated at 1 300 kronor per day. This represented a saving of about 45 000 kronor per year.

In terms of the costs and benefits of the investment itself, Alternative I suggests that the safety analysis and control measures resulted in expenses of 725 000 kronor being incurred. However, under Alternative II the investment was estimated to represent an immediate saving of over 2 million kronor.

In operational terms, the flow of future earnings can be calculated as a current capital value (using the discounting method described in sect. 13.4). Assuming an investment life of 10 years and a discount rate of

Table 15.1 *Summary of the costs (–) and benefits (+) of a safety analysis applied to the rebuilding of part of a paper mill (SEK 000)*

Activity	Alternative I	Alternative II
INVESTMENTS		
Safety analysis (net)	– 75	– 75
Investment costs		
Layout and transportation system	– 900	– 900
New paper machine	0	– 200
Remodelling of existing paper machine	– 200	– 200
Investment benefits		
Layout and transportation system	+ 350	+ 3 500
Remodelling of existing paper machine	+ 100	+ 100
Investment benefit/cost	– 725	+ 2 225
OPERATIONS		
Improved productivity	+ 270	+ 270
Reduced risk for disturbances	0	+ 85
Fewer accidents	+ 45	+ 45
Annual operational benefit/cost	+ 315	+ 400

10%, the current value of future earnings becomes 1.9 or 2.4 million depending on the alternative chosen. The more careful form of planning and design that safety analysis involves would have been profitable whichever method of appraisal was adopted. However, it would not have been profitable if only the savings from a reduced number of accidents had been taken into account.

ANALYSIS OF LAYOUT AND THE MATERIALS TRANSPORTATION SYSTEM

In conjunction with the rebuilding of the paper mill, a major change was made to the materials transportation system. Moreover, several of the machines were repositioned. The conveying of materials to and from the rolling/cutting machines and the storing of finished products form an important part of operations at a paper mill. It is governed from one end by the paper-making machinery, from the other by the need to make deliveries on time.

Layout and transportation were analyzed together. The analyses provided part of the basis for making decisions on the positioning of machines, walls, doors, etc. They also applied to how fixed conveyors should be designed.

The issues were discussed at six special meetings of the work environment group, each of which lasted for about two hours. The discussion was based on a proposal for layout and materials transportation. The proposed arrangement was divided up into suitable sections, each of which were analyzed separately. Energy analysis and Deviation analysis were employed simultaneously.

The analyses focused directly on immediate control measures: when a problem was identified, the aim was to find a direct solution. To summarize all hazards in one go, and then take all necessary measures together would have been an unmanageable procedure.

The analysis was conducted in the following stages:

1. The normal materials flow was checked. Energies and the deviations and problems that might arise were studied for each section.

2. Occasional forms of materials transportation were analyzed separately. These analyses were applied, for example, to small rolls that were liable to tip, to a number of supplementary materials, and to

waste disposal. These work phases tended to involve various types of manual handling.

3. Pedestrian traffic was treated as a special phase of work. Different types of hazards were identified, resulting in a proposal for suitable walkways.

4. Special attention was paid to industrial truck traffic. Trucks were used for both routine transportation and occasional forms of materials conveyance.

5. Maintenance and repairs were affected by the design of the layout. Specific points investigated were accessibility, lifting facilities and parts transportation.

Documentation was prepared by directly writing in changes on a drawing. When this document was redrawn, it provided the basis for a new round of analyses.

Certain parts of the layout were not covered by the analyses due to a lack of time. This depended in turn on changes having being introduced, partly as a result of earlier analyses. It meant that certain parts

Some occasional forms of materials transportation were covered by the analysis. Here, waste paper from a machine is being disposed of

of the transportation system were ordered without there having been an opportunity for the study team to inspect them. It emerged later that some of the hazards that remained were related to these changes (see also sect. 14.4 and table 14.2).

Economic appraisal

The analyses and discussions led to improvements and changes, some of which were extensive. The most expensive concerned a conveyor belt system. It was decided that the belt should be extended and repositioned at ground level, which proved to be complicated to implement in practice. In addition, there were general construction expenses. A total cost of around 900 000 kronor was incurred.

Savings could be made by simplifying the system in several places, and these were valued at 350 000 kronor. It was originally intended that automatic industrial trucks should form part of the transportation system. This idea was abandoned at the planning stage, resulting in a saving of 3.5 million kronor. Safety considerations partly lay behind this decision.

The analyses gave rise to extra design work (of about a week), but this was compensated for to a certain extent by coordination benefits. The length of the run-in period was not affected, either positively or negatively.

The changes did not lead to any increased operational expenses. The changed layout provided more storage space, which was valued at 180 000 kronor per year. The cost of transportation by truck was reduced by 90 000 kronor per year through the creation of a more rational materials flow than that contained in the old proposal. In total, therefore, earnings to a value of around 270 000 kronor a year accrued.

The risks for disturbances to production may well have been reduced. Although such gains are difficult to estimate, an example is provided here. The new layout offered better opportunities for work to be carried out even when the fixed conveyance system was out of order. A breakdown in the materials handling system might halt production for 24 hours — at a cost of 1.5 million kronor. On the assumption that this could be avoided once every 20th year, the reduced risk can be estimated to have a monetary value of 85 000 kronor per year. Modification of an overhead crane has reduced the risk of handling injuries. A hypothetical value of 20 000 kronor can be attributed to this — on the grounds that one minor injury can be avoided every other year.

PURCHASE OF A NEW ROLLING MACHINE

A new paper rolling machine was to be purchased while the section was being rebuilt. The idea from the outset was that this should be as safe as possible. The analysis was based on offers made by two machine suppliers and the study of two similar machines already in operation at other paper mills.

The two machines were analyzed using Energy and Job safety analysis. Study visits were made to gather information on accidents and near-accidents, disturbances to production and maintenance problems. These resulted in a requirement specification for safety, maintenance, etc., which was then presented to and accepted by both manufacturers. The specification was later included in the purchasing contract.

The eventual supplier originally required an extra 350 000 kronor to meet the new work environmental requirements. About half of this amount concerned safety, the other half noise and vibrations. This extra amount was negotiated away, and the final price of the machine was not affected. Thus, as the extended specification was not estimated to provide any operational benefits or costs, the net cost/benefit of the analysis might reasonably be regarded as zero. On the other hand, without the new requirements, it might have been possible to reduce the purchase price by a further 200 000 kronor. This is the figure employed in the alternative estimate.

REMODELLING AN EXISTING MACHINE

An existing paper rolling machine was to be moved to a new position during rebuilding. It was originally intended that there were to be no changes to the machine, but a number were made for safety reasons.

The analyses were carried out by a study team consisting of three people. Job safety analysis was the principal method employed. This covered 36 phases of normal job procedure plus a few extra phases to account for occasional tasks. The analysis took about 5 minutes per phase, and was then supplemented by Deviation and Energy analysis.

A special study was made of five job phases where there was a particularly serious risk of a part of the body getting caught between paper rolls and reel cutters. It proved possible to find alternative work procedures which were expected to considerably reduce the risks.

The hazards at this paper machine were assessed to be sufficiently great for control action to be necessary. The analyses resulted in a list of safety measures containing around 50 items. These could be divided up into the following categories:

— Technical arrangements.

— Control systems for the paper rolling machine.

— Job procedures.

— Inspections (to be carried out more regularly).

— Functional requirements (not specified in detail).

A particularly hazardous phase of work is to feed in paper from a roll that is already in position

Financial appraisal

The remodelling of control and hydraulic systems cost around 200 000 kronor. However, the same sort of conversion would have been needed sooner or later. An extra cost of 100 000 kronor was incurred by having to transport rolls to another paper mill while the conversion took place. This could have been avoided if the work had taken place at a more appropriate time.

236

Operational costs have not been affected. Some employees have maintained that the work takes a bit longer because certain shortcuts are no longer permissible; others think that the finished rolls are of better quality.

15.3 Purchase of packaging equipment for paper rolls

Equipment for the automatic wrapping of paper rolls was purchased. There was assessed to be a need for accident risks to be thoroughly examined. A request was made for a safety analysis to be conducted. The results of the analysis would then be discussed with the supplier and taken into account when the system was designed and installed (Backström & Harms-Ringdahl, 1986; Harms-Ringdahl, 1986).

The planned installation was complicated. It was controlled by a computer which needed to be coordinated with other computerized systems. About 30 mechanical movements were involved and two materials handling robots were included. To some extent this was a tailor-made system, but a number of similar systems had previously been manufactured.

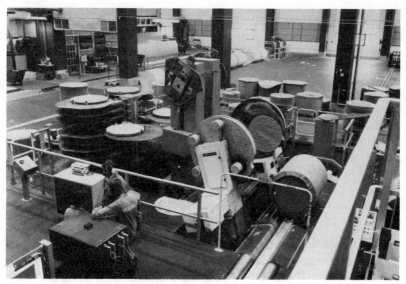

An overall view of the packaging installation

The amount of time available for the analysis was limited and not really sufficient for an installation of such size and complexity. Study visits, each of just over a day, were made to two similar plants. Including written reporting and meetings, the analysis took just over a week.

Energy and Deviation analysis were applied. In the latter case, there was insufficient time for the analytical procedure to be rigourously followed through. An attempt was made to identify as many deviations as possible through interviews with the people involved. There were a large number of powerful machine movements. Also, there were several examples of disturbances to production whose correction required work in proximity to the machines. The safety analysis resulted in a list of hazards and ideas for safety measures. This list was employed in discussions between the paper mill and the supplier.

Economic appraisal

Just over a week was devoted to the analysis. Had it not been conducted, other consultancy services would have been needed. At the paper mill, the judgement was that the safety analysis was twice as efficient as a more common form of investigation would have been.

In addition, the safety analysis had an educational effect equivalent to one man week of training. Thus, the analysis itself generated a net labour saving of just over two weeks, with an estimated value of 30 000 kronor.

The changes have not given rise to either extra expenses or time delays for the customer. The run-in period was considered to have been reduced by a month through the identification of a number of problems in advance. The putting into operation of similar machines from the same supplier had been delayed by several months as a result of run-in problems. A simple calculation can be made on the basis of the value of the installation, assessed at 35 million. Assuming that this is depreciated over 10 years and that the equipment is available for one extra month, a figure of around 300 000 kronor is obtained. Perhaps half of this amount can be ascribed to the safety analysis.

That demands for improvements arose before the machine itself was manufactured may be expected to have provided further savings, mostly for the supplier. Those accruing to the customer are valued at 70 000 kronor.

The effect on operations is difficult to assess. In the judgement of the supplier, the careful work conducted by the paper mill at the planning stage has provided for better machine accessibility than that for other machines. Employing cautious reasoning, economic costs and benefits might both be assessed at zero. Also, it is hard at present to estimate how the risk for accidents has been affected. It is difficult to place a value on either this or the part played by the safety analysis. For this reason, it was decided to set the value of these possible savings at zero.

In the course of the analysis, certain problems related to man-machine interaction were observed. This has led to improvements at the work-place, probably resulting in rolls of higher quality and better appearance. A theoretical value can be allocated to this. Defective or poorly-wrapped rolls can lead to an order being missed, one perhaps in an amount of 15 millions. Even supposing that the improvement leads to one fewer missing order every 100 years, the annual saving will be 150 000 kronor per year.

The economic considerations are summarized in table 15.2. The cautious estimates of Alternative I result in a zero financial outcome: the safety analysis neither incurs costs nor generates income.

Table 15.2 *Summary of the costs (−) and benefits (+) of a safety analysis applied to a packaging station at a paper mill (SEK 000)*

Activity	Alternative I	Alternative II
INVESTMENTS		
Safety analysis (net)	0	+ 30
Investment costs	0	0
Investment benefits		
Shorter run-in time	0	+ 150
More rational planning and design	0	+ 70
Investment benefit/cost	0	+ 250
OPERATIONS		
Improved quality, etc.	0	+ 150
Fewer accidents	0	0
Annual operational benefit/cost	0	+ 150

Under Alternative II, analysis at the planning stage results in an investment-related saving of 250 000 kronor. The theoretical annual gains from operations are 150 000 kronor. Under the assumptions of an investment life of 10 years and a discount rate of 10%, this has a current capital value of 0.9 million. The total profit comes to 1.15 million kronor.

15.4 Examination of a computer-controlled materials handling system

A study has been conducted of an installation for the automated sorting of loading pallets. The pallets are moved into a sorting area and inspected automatically. The ones that are damaged are sorted to one side to be sent on to the repair shop. Those that are wholly intact are stacked, fastened and stored.

The equipment is tailor-made and designed for fully automated operation. It is controlled by a computer which governs fifty different mechanical movements. Information comes from about a hundred sensors.

Part of the automated pallet handling system

The installation was used as a practical example in a training course on safety analysis. At this course, around thirty proposals for safety improvements were made. Before this, normal inspections of the equipment had been carried out by a safety engineer and an officer from the Labour Inspectorate, both with considerable experience. At these inspections, six accident-related proposals had been made. This suggests that safety analysis can be a significantly more efficient means of identifying occupational hazards.

A deeper analysis of the installation was made in conjunction with a later research project (Backström & Harms-Ringdahl, 1986; Harms-Ringdahl, 1986). This led to the generation of further safety proposals. In short, the analyses demonstrated that there were a number of machine movements for which protective devices were lacking. One explanation for this was the assumption made at the design stage that the machines would run wholly automatically. Accordingly, it was believed that there would be no reason for anyone to be in proximity to them.

There proved, however, to be a large number of reasons why manual interventions had to be made. There were several possible sources of disturbances to operations, all of which required manual correction. The safety proposals focused on how these should be handled and, to a certain extent, on how the frequency of such disturbances should be reduced.

Economic appraisal

The appraisal below is based on an imaginary situation, i.e. that a safety analysis had been conducted at the planning and design stage. No distinction has been made between supplier's and customer's costs as the distribution of these would have been a matter for negotiation between them. Although the reasoning employed in this example is rather complicated, it may give the reader an idea on how to proceed in similar situations.

It can be estimated that a couple of weeks would have been necessary for the analysis and discussion of safety measures, at an estimated cost of 40 000 kronor. Further expenses would have arisen from extra design and construction work, which may be estimated to have cost between 80 000 and 150 000 kronor.

It may have been possible to implement the project without any significant delay. An alternative assumption is that system start would have been delayed by three months. This would have meant that four employees would have had to continue with the manual handling of pallets for this period — at an estimated cost of 300 000 kronor.

The run-in time for the installation was fairly long. It should have been possible to make it considerably shorter with more careful planning and design — providing a conceivable saving of 70 000 kronor. Some changes to the equipment have been made since it was put into operation. These could have been avoided, resulting in a saving of 40 000 kronor.

On the operational side, the frequency of disturbances has meant that one person has to monitor the equipment virtually all the time. The work station for the person doing the monitoring is poorly designed. Had it been better planned, work conditions would have been better and efficiency greater. It was estimated that 20% of working time was lost as a result of the poor design of the work station. It would have been possible to utilize this time for other activities, and this has been attributed a value of 60 000 kronor as "Improved efficiency" under Alternative I. If some of these disturbances had been prevented, it might have been possible to make savings equivalent to half the working time of one man. There are also opportunities for this time to be utilized for other tasks at the plant. Accordingly, the labour cost of a half-time worker (150 000 kronor) has been included under Alternative II.

The installation has a certain amount of over-capacity, but overtime working is still needed to compensate for interruptions to production. The cost of this has been estimated at 40 000 kronor per year. It should have been possible to prevent a large number of the halts to production, so these savings have been included under the heading "Increased reliability".

So far, no accidents leading to an absence from work have occurred, which suggests that increased safety has a low monetary value.

Table 15.3 lists the various items included in the calculation. Assuming an investment life of 10 years and a discount rate of 10%, the current value of future earnings is 0.6 or 1.2 million kronor depending on the alternative chosen. On both estimates, it would have been of economic benefit to have conducted a safety analysis before the project was implemented.

Table 15.3 *Summary of the costs (−) and benefits (+) of a hypothetical safety analysis applied to a computer-controlled materials handling system (SEK 000)*

Activity	Alternative I	Alternative II
INVESTMENTS		
Safety analysis (net)	− 40	− 40
Investment costs		
Design and construction	− 80	− 150
System delay	0	− 300
Investment benefits		
Shorter run-in time	0	+ 70
Avoidable changes	+ 40	+ 40
Investment benefit/cost	− 80	− 380
OPERATIONS		
Improved efficiency	+ 60	+ 150
Increased reliability	+ 40	+ 40
Fewer accidents	0	0
Annual operational benefit/cost	+ 100	+ 190

15.5 Examination of a section for the mixing of chemicals

A safety engineer was requested to come up with proposals to reduce high levels of absence through sickness and personnel turnover at a unit for the mixing of chemicals (Harms-Ringdahl, 1991). The chemicals were mixed for the production of ceramic materials. A lot of manual work was involved, some of which was heavy. Moreover, a large amount of quartz dust was generated by the work. The safety engineer had conducted an investigation and made proposals a few years earlier, but these were rejected by the company on the grounds that they were too expensive.

On this occasion, the safety engineer carried out a safety analysis and supplemented the analysis with an economic appraisal. Energy analysis and Deviation analysis were employed. The entire investigation took about a week, of which one day was devoted directly to the analysis itself.

The analysis revealed there were many reasons why accidents might occur. In addition, there were several possible sources of human error which would result in a faulty mixture. When this was followed up, it emerged that such errors had been made several times a year but that production management was generally not aware of these. When one was detected, the mixture was just disposed of. Usually, it was simply poured down a drain, also creating an environmental problem. If the error was not detected, a faulty mixture was introduced into the production process, leading to major disturbances to production at a later stage. The costs are difficult to estimate, but they have been allocated a value of 1 million kronor per year in the calculation.

The estimates are summarized in table 15.4. Assuming an investment life of 10 years and a discount rate of 10%, the current value of future earnings sums to 2.2 or 8.3 million kronor depending on the alternative chosen. There would be clear economic advantages to implementing the changes suggested by the safety analysis.

When company management was made aware of the appraisal (using

Table 15.4 Summary of the costs (−) and benefits (+) of a safety analysis applied to the conversion of a production unit for the mixing of chemicals (SEK 000)

Activity	Alternative I	Alternative II
INVESTMENTS		
Safety analysis (net)	− 20	− 20
Investment costs		
Technical changes	− 200	− 200
Investment benefits	0	0
Investment benefit/cost	− 220	− 220
OPERATIONS		
Overexertion injuries	+ 40	+ 40
Absence	+ 50	+ 50
Personnel turnover	+ 70	+ 70
Wastage of materials	+ 200	+ 200
Disturbances to production	0	+ 1 000
Annual operational benefit/cost	+ 360	+ 1 360

Alternative 1), an immediate decision was made to reconstruct the mixing facilities.

15.6 Control of a line of sheet-metal presses

A production line containing several sheet-metal presses was about to be installed. Some parts for the line required detail design. The safety engineer at the company had not previously been involved in the project. He was invited to a meeting with the supplier and project leader to assess possible hazards before the design of these parts was finalized.

The safety engineer proposed that a simple safety analysis should be conducted. His suggestion met with a favourable reception. He started by presenting the principles of Energy and Deviation analysis for half an hour. A tour was then made of the site and an immediate "quick" analysis conducted. This took one and a half hours. Drawings of the design proposals were used to supplement the analysis. The group reassembled and went through their notes.

Together they found thirty or so mechanical hazards, and the proposals principally concerned supplementing the line with machine guards, railings and warning equipment. The costs of the measures were calculated, and an immediate decision on implementation was taken at the meeting.

During the discussion the safety engineer posed a number of questions concerning the computer-based control system which none of the participants could answer. There was a whole series of problems on which nobody had a real grip. These problems were later raised with a consultant. The methodological perspective provided by safety analysis enabled the safety engineer to engage in meaningful dialogue with the consultant, despite the fact that the control system lay outside his own area of expertise.

15.7 Integrated analysis of a chemicals plant

The management of a chemicals processing plant wanted a summary risk analysis to be conducted, with the aim of obtaining an overall picture of hazards. In addition, an assessment was required of whether Swedish regulations based on the Seveso directive (CEC, 1982) applied to the company's operations.

The level of ambition of the analysis was determined by the requirement that the work should be conducted in one week. This meant that some form of quick or rough analysis had to be conducted. This was carried out in the form of two meetings of the study team, each lasting around four hours.

The first meeting

It was decided at the first meeting that the entire installation should be examined. To start with, all participants noted down the serious hazards of which they were aware. A list of 10 in total was obtained. A quick form of analysis based on a combination of the principles of Energy analysis and Deviation analysis was chosen. The object was structured using a rough block diagram. An integrated approach was chosen, and the types of consequences to which attention should be paid were:

— Fires.

— Explosions.

— Serious occupational accidents.

— Damage to the neighbouring environment.

— Disturbances to production.

Around 100 different hazards were identified.

The second meeting

Risks were then classified as follows:

I   Acceptable for the time being

II   Measures to be taken as soon as possible.

III   Immediate measures essential.

A   Investigation or safety analysis recommended

In some cases it was possible to estimate the economic consequences of an accident. Twelve hazards were picked out and safety proposals were produced in relation to these.

Results

Around 100 different hazards were identified and about 30 safety proposals made, some of which were general by nature. The most severe accident that might reasonably be expected to occur would kill five people. This referred to a fire in a distillation column for solvents. Such an accident might be triggered off by an electrical power failure, which would cause overheating and an explosion if corrective measures were not taken immediately. A request was made to the National Board of Occupational Safety and Health for an interpretation of whether the Seveso Directive was applicable to this situation. The Board replied that the directive did not apply — as the consequences of such an accident were not sufficiently great.

Comments

A simple analytical method was employed but a large number of hazards were identified and proposals made. The fruitfulness of the analysis is related to the study team's considerable knowledge of the plant and the risks involved. The role of the analysis was to stimulate the imagination of the team members, summarize their joint knowledge, and then present it in a systematic manner.

15.8 Accident investigation

A field trial

Experimental safety routines were introduced at two paper mills for a period of around six months. Accidents were investigated more thoroughly and systematically than had hitherto been the case (Harms-Ringdahl, 1983; 1987b; 1990). The trial was conducted according to the principles discussed in sections 6.5 and 13.2. An investigation group consisting of four people was set up at each mill. These were the safety engineer, the safety technician, the trade union safety representative and the production section manager.

It had been usual procedure for a simple accident investigation to be carried out by the job supervisor responsible for the area where the accident took place. The report was then forwarded for registration and entry into the accident statistics. In some cases, the occurrence of an accident had led to the taking of safety measures.

A total of 12 accidents and 3 near-accidents were investigated in the course of the trial, making a total of 15 cases. The investigations themselves took about 30 minutes per case. Including meetings and preparing reports on safety measures, etc., the total time devoted to each case came to around one hour.

Results

The quality of investigations was improved. In comparison with previous investigations:

— The amount of information on events preceding an accident (number of deviations) was doubled.

— Five times as many safety measures were proposed.

At paper mills, working with rolling machines is particularly hazardous. The frequency of accidents is well above the average for the paper industry. During the experimental period, priority was given to accidents at these machines and a total of seven accidents was investigated.

The result was that the number of accidents at paper rolling machines fell by a third. The number of days of sick leave resulting from such ac-

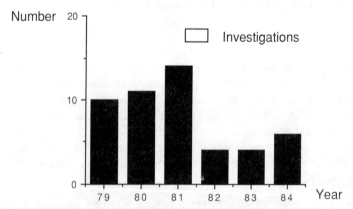

Figure 15.2 *Accidents at paper rolling machines in two mills before and after the field experiment*

cidents also fell — from an annual average of 420 to 130 days. There was also a reduction in the frequency of accidents occurring in the course of other activities at the paper mill, but this cannot be attributed with the same degree of certainty to the accident investigation routine.

Economic appraisal

The trial activities took up working time of about two man weeks in total. Those who took part considered that the educational value of participation could be valued at about two days training per partici- pant, i.e. just over one man week. The time used was created by the re- distribution of job tasks, which meant that no net cost was incurred. However, help was provided for training and getting the experiment un- derway. Let us value this at 20 000 kronor.

The economic appraisal is restricted to the paper rolling machines, where the effects are most easily discernible. Half of the time spent on experimental activities was devoted to these machines. Technical mea- sures cost about 30 000 kronor. The greatest effort concerned a thor- ough examination of the control system for one of the machines. This had caused problems and had led to both accidents and near-accidents. Solving these problems took up four man weeks, but such an examina- tion would have had to have been carried out sooner or later. Under Alternative II, the cost for this was estimated at 40 000 kronor.

Table 15.5 Summary of the costs (–) and benefits (+) of an accident in- vestigation routine (SEK 000)

Activity	Alternative I	Alternative II
INVESTMENTS		
Safety analysis (net)	– 20	– 20
Investment costs		
Technical changes	– 30	– 30
Examination of control system	0	– 40
Investment benefit/cost	– 50	– 90
OPERATIONS		
Fewer accidents	+ 36	+ 225
Annual operational benefit/cost	+ 36	+ 225

After the experiment, the number of accidents fell by an average of eight a year, and there were 300 fewer days of sick leave as a result. The company estimated an accident to result in an average cost of around 4 500 kronor. Another conceivable way of calculating this cost is to attribute 750 kronor to each day of sick leave taken (Alternative II).

In this case, earnings have been calculated over a period of three years, as major changes were then made to the installation. A discount rate of 10% has been assumed. The current value of the future income attributable to the investment can be estimated at 90 000 or 560 000 kronor depending on the alternative chosen. The safety analysis was profitable even in this case, despite the fact that the "only" gain came from the occurrence of fewer accidents.

Comments

Examples of problems involved in the experiment were:

— The production section managers attended only a few of the meetings.

— The investigations took place a long time after the accidents occurred (40 days on average instead of 3 days as planned at the beginning).

— It was sometimes difficult to get safety proposals implemented.

It was intended that the trial should last for a period of six months, and it ended immediately after that period expired. A decision was made not to reinstate the routines. This decision was not changed even when the results had demonstrated a considerable reduction in the number of accidents. The explanation presented to the author was that the company wanted to wait for the reaction of the paper industry work environmental council. As no reaction was forthcoming, activities were not recommenced.

In fact, it is surprising that the fall in the number of accidents was so great and that the lower accident frequency was maintained throughout the three years for which evaluation was possible. The results can only be partly explained by the control measures taken. No study of possible explanations has been conducted. Nevertheless, comments made at the end of the experimental period hint at some conceivable reasons:

"Now we have a far better language with which to discuss accidents."

"What was difficult was that supervisors became irritated when asked

about accidents. Their view was that they had carried out an investigation and everything had been cleared up."

The first comment refers to people beginning to apply the ideas that lie behind Deviation analysis, e.g. that there is always a pre-history to the occurrence of an accident.

The second refers to the interest in accidents that had been generated by the experiment. Previously, an accident investigation had been regarded as a piece of paper that had to be sent off. During the experiment an investigator arrived and asked questions. It may be supposed that this initiated a process through which greater attention was paid to accidents. Attitudes towards them probably changed, above all among job supervisors. This may well be the most important reason why the number of accidents fell so dramatically.

Accident investigation is not a form of safety analysis in a strict sense, as only part of the system and just a few hazards are covered. The "selection" of cases is made at random, i.e. by accidents that have occurred. Deviation theory suggests that "organizational deviations" lie behind most accidents. Such deviations can be rather general by nature, which explains why investigations can provide a type of knowledge which applies to the whole installation.

15.9 Development of safety rounds

There was discontentment with the way in which safety rounds were conducted at a company. The custom was that a round should take place between once every fourteen days and once a month, depending on the nature of activities. A total of 15 safety areas was involved. One of the problems was that measures decided upon were not implemented, with the result that the safety representative found it frustrating to continue to point out deficiencies in the work environment. The issue of dissatisfaction with safety rounds was taken up at a meeting of the safety committee. The safety engineer, in conjunction with a study team, was allotted the task of proposing how these activities could be improved.

At the meeting, the safety engineer pointed to a safety analysis that had been carried out at one of the company's plants. It resulted in attention being drawn to fifty or so defects (see sect. 15.5). As no changes had been made to this section of the factory for the previous ten years, people should have already been aware of these defects. At the same time, a rough calculation was made of how many working hours had been devoted to safety rounds over this period. In the light of the fact that the defects had not been detected on these rounds, the safety engineer offered the opinion that resources should be diverted to more concrete measures. As a result, a hypothesis was suggested — that the company's safety rounds were ineffective.

Against this background, a questionnaire was sent to all job supervisors and trade union safety representatives. Questions concerned how often safety rounds were conducted, who participated, and how it was felt they operated.

In conjunction with the survey, a spot check was made. To the question whether industrial truck charging stations were usually inspected during the round, both the supervisor and the safety representative had responded that they always were. They also reported that they had "adequate" or "quite good" knowledge in this area. Together with the chief safety representative and maintenance manager, the safety engineer then went round the plant's 15 truck charging stations. The tour revealed the existence of just over 60 defects, some of which were in direct breach of safety directives. The lack of correspondence between the questionnaire responses and the defects discovered spoke clearly of the gulf between belief and reality. It also supported the hypothesis that safety rounds were ineffective.

The material was presented to the safety committee. The conclusion was drawn that the safety rounds were not of sufficient quality. It was decided that the number of rounds should be cut down. Instead, there should be targeted safety rounds, each with a specific theme, which should take place just a few times a year. Before each round, the participants should receive training on the theme in question.

Just over a year later, five such rounds had been implemented. The safety engineer supported activities by providing training, and taking an active part in the new routine. The themes were as follows:

— Cleaning — methods and technical conditions, especially those concerned with reducing quartz dust.

— Ergonomics.

— Working conditions of cleaners.

— Machinery lacking nip guards, positioning of emergency stop buttons.

— Emergency escape routes — design and signposting.

These "thematic" rounds provided an opportunity for participants to be introduced to a simple form of risk analysis and also supplied occasions on which it could be applied. This offered new perspectives on how the work environment could be scrutinized, compared with the traditional way of identifying hazards and evaluating risks.

16. Concluding remarks

Safety analysis is by no means an established way of achieving occupational safety. For this reason, this book should in part be regarded as a source of ideas. The area will certainly be further developed. The methods described have a general character. One possible way in which safety analysis might be developed is through the creation of more specialized methods, e.g. those that are specific to a certain type of installation or piece of equipment.

Some of the examples concern a type of chemicals plant. This does not mean that the use of these analyses is limited to this industry. The reason for employing a chemicals installation as an example is that it is fairly easy to describe as an object. Other types of installations often need to be described in far greater detail to make it possible to understand how they operate.

I would like to stress one final time that analyses are best conducted in a team. The leader of the analysis must work together with people who know the system, thus allowing him or her to concentrate on the analysis itself. It is further recommended both that an integrated approach to safety analysis is adopted and that economic arguments (which are often very powerful) should be employed to back up safety proposals. Safety is not always an expense. It can be good business as well.

The focus of this book has been on occupational accidents and what can be done about them at company level. But the principles should also be applicable to a broader spectrum of systems and undesired events. It would certainly be of advantage if experiences and knowledge of applications could be exchanged across a wider area.

The scope of this guide is fairly extensive, as it is designed to cover several methods and different areas of application. Using safety analysis can appear demanding at first acquaintance. But it need not be particularly difficult in practice. In many cases, extensive planning is not required and it is enough to be familiar with just a few methods.

It is probably only when you have conducted a safety analysis yourself that you recognize the benefits of this way of working. You detect hazards that would otherwise have remained undiscovered. This can be a stimulating experience.

17 References

Andersson, R. The role of accidentology in occupational injury research. National Institute of Occupational Health, Solna, Sweden, 1991.

Amendola, A., Contini, S., & Ziomas, I. Uncertainties in chemical risk assessment; Results of a European benchmark exercise. *Journal of Hazardous Materials*, **29**, 347–363, 1992.

Association of Swedish Chemical Industries. Säkerhetsgranskning. En vägledning för skadeförebyggande och skadebegränsande arbete vid industriell kemikaliehantering. Kemikontorets förlags AB, Stockholm, 1985.

Backström, T. & Harms-Ringdahl, L. A statistical study of control systems and accidents at work. *Journal of Occupational Accidents*, **6**, 201–210, 1984.

Backström, T. & Harms-Ringdahl, L. Mot säkrare styrsystem — Om personsäkerhet vid automatiserad produktion. Royal Institute of Technology, Stockholm, 1986.

Bainbridge, L. The ironies of automation. In *New technology and human errors*, ed. J. Rasmusen, K. Duncan & J. Leplat. Wiley, London, 1987.

Bartlett, F.C. *Remembering: A study in experimental and social psychology.* Cambridge University Press, Cambridge, 1932.

Bell, R., Daniels, B.K. & Wright, R.I. Assessment of industrial programmable electronic systems with particular reference to robotics safety. Fourth National Reliability Conference — Reliability 83, England, 1983.

Bell, J.B. & Swain, A.D. A procedure for conducting a human reliability analysis for nuclear power plants. U.S. Nuclear Regulatory Commission, Washington, 1983.

Bergman, B. On reliability theory and its applications (with discussion). *Scandinavian Journal of Statistics*, **12**, 1–41, 1985.

Bird, F.B. & Loftus, R.G. *Loss control management.* Institute Press, Loganville, Georgia, 1976.

Brehmer, B. The psychology of risk. In *Risk and decisions.* ed. W.T. Singleton & J. Hovden. Wiley, 1987.

Brehmer, B. Distributed decision making: Some notes on the literature. In *Distributed decision making. Cognitive models for cooperative work.* ed. J. Rasmussen, B. Brehmer & J. Leplat. Wiley, 1991

Britter, R.B. The evaluation of technical models used for major-accident hazard installations. University of Cambridge, Cambridge, 1991.

Brody, B., Letourneau, Y. & Poirier, A. An indirect cost theory of work accident prevention. *Journal of Occupational Accidents*, **13**, 225–270, 1990

Brown, D.M. & Ball, P,W. A simple method for the approximate evaluation of fault trees. 3rd International Symposium on Loss Prevention and Safety Promotion in the Process Industries, European Federation of Chemical Engineering, 1980.

Bullock, M.G. Change control and analysis. EG&G Idaho Inc, Idaho, 1976.

Burström, L. & Lindgren, G. Modell för lokalt skyddsarbete, University of Luleå, Luleå, Sweden, 1983.

Contini, S., Amendola, A. & Ziomas, I. Benchmark Exercise on Major Hazard Analysis. Commission of the European Communities, Joint Research Centre, Ispra, Italy, 1991.

CEC, Council of the European Communities. Council directive on the major-accident hazards of certain industrial activities. *Official Journal of the European Communities*, 1982.

Center for Chemical Process Safety. *Guidelines for chemical process quantitative risk analysis*. American Institute of Chemical Engineers, New York, 1989.

CISHC, Chemical Industry Safety and Health Council. *A guide to Hazard and operability studies*. Chemical Industries Association, London, 1977.

Clemens, P.E. A compendium of hazard identification and evaluation techniques for system safety application. *Hazard Prevention*, **18**, 11–18, 1982.

Committee on Trauma Research. *Injury in America*. National Academy Press, Washington, 1985.

DiNenno, P.J. (ed.) *SFPE Handbook of fire protection engineering*. National Fire Protection Association, Quincy, Maryland, 1988.

Dow Chemical Company. Fire and Explosion Index. Hazard Classification Guide. 4th ed. Midland, Michigan, 1976.

Döös, M. & Backström, T. Lär av olycksfallen; en utredningsmodell med exemplifierande resultat. National Institute of Occupational Health, Solna, Sweden, 1990.

Farmer, E. & Chambers, E. A psychological study of individual differences in accident rates. Industrial Health Research Board, report No 38, London, 1926.

Ferry, T.S. *Modern accident investigation and analysis*. Wiley, N.Y., 1981.

Fischoff, B., Lichtenstein, S., Slovic, P., Derby, L. & Keeney, R. *Acceptable risk*. Cambridge University Press, Cambridge, 1981.

Freud, S. *Psychopathology of everyday life*. Ernest Benn, London, 1914.

Fussell, J.B. Synthetic fault tree model. A formal methodology for fault tree construction. Aerojet Nuclear Company, Nat. Reactor Testing Station, Idaho, 1973.

Fussel, J.B. Fault tree analysis: concepts in techniques. In *Generic Techniques in systems reliability assessment*, ed. E. J. Henley & J. W. Kynn, J. W. Noordhoff, Leyden, 1976.

Fång, G. Huru många arbetsolyckor inträffa egentligen? *Arbete, Människa, Miljö*, **1**, 73–76, 1988.

Fång, G. & Harms-Ringdahl, L. Safety considerations in the design of factories — a study of three cases. *Journal of Occupational Accidents*, 6, 1984.

Gibson, J.J. Contribution of experimental psychology to the formulation of the problem of safety: a brief for basic research. In *Behaviour Approaches to Accident Research*. Association for the Aid of Crippled Children, New York, 77–89, 1961.

Gordon, J.E. The epidemiology of accidents. *American Journal of Public Health,* **9,** 504–515, 1949.

Gossens, L.H.J. Risk prevention and policy-making in automatic systems. *Risk Analysis,* **11,** 217–228, 1991.

Grimaldi, J. Paper at the ASME standing committee on Safety. Atlantic city, N.J., 1947.

Haddon, W. Jr. A note concerning accident theory and research with special reference to motor vehicle accidents. *Annals of New York Academy of Science,* **107,** 635–646, 1963.

Haddon, W. Jr. The basic strategies for reducing damage from hazards of all kinds. *Hazard Prevention* **16,** 8–12, 1980.

Hale, A.R. Safety Rules O.K.? Possibilities and limitations in behavioural safety strategies. *Journal of Occupational Accidents,* **12,** 3–20, 1990.

Hale, A. & Glendon, I. *Individual behaviour in the control of danger,* Elsevier, Amsterdam, 1987.

Hammer, W. *Handbook of system and product safety.* Prentice Hall Inc., New Jersey, 1972.

Hammer, W. *Product safety management and engineering.* Prentice Hall, Inc., New Jersey, 1980.

Harms-Ringdahl, L. Riskanalys vid projektering — Försöksverksamhet vid ett pappersbruk. Royal Institute of Technology, Stockholm, 1982.

Harms-Ringdahl, L. Vis av skadan — Försök att systematiskt utreda och förebygga olycksfall vid två pappersbruk. Royal Institute of Technology, Stockholm, 1983.

Harms-Ringdahl, L. Experiences from safety analysis of automatic equipment. *Journal of Occupational Accidents,* **8,** 139–146, 1986.

Harms-Ringdahl, L. Safety analysis in design — evaluation of a case study. *Accident Analysis and Prevention,* **19,** 305–317, 1987*a.*

Harms-Ringdahl, L. *Säkerhetsanalys i skyddsarbetet — En handledning.* Folksam, Stockholm 1987*b.* (Original version of this book.)

Harms-Ringdahl, L. On economic evaluation of systematic safety work at companies. *Journal of Occupational Accidents,* **12,** 89–98, 1990.

Harms-Ringdahl, L. *Att förebygga olycksfall — Säkerhetsanalys i företagshälsovården.* Swedish Work Environment Fund, Stockholm, 1991.

Health and Safety Executive. *Human factors in industrial safety.* Health and Safety Executive, London, 1989.

Health and Safety Executive. *Successful health & safety management.* Health and Safety Executive, London, 1991.

Heino, P., Poucet, A. & Suokas, J. Computer tools for hazard identification, modelling and analysis. *Journal of Hazardous Materials,* **29,** 445–463, 1992.

Heinrich, H.W. *Industrial Accident Prevention.* McGraw-Hill, New York, 1931.

Heinrich. H.W., Petersen, D. & Roos, N. *Industrial Accident Prevention.* 5th ed. McGraw-Hill, New York, 1980.

Hendrick, K. & Benner, L. *Investigating accidents with STEP*. Marcel Dekker, New York, 1987.

Henley, J. H. & Kumamoto, H. *Reliability engineering and risk assessment*. Prentice-Hall Inc., New Jersey, 1981.

IEC, International Electrotechnical Commission. Analysis techniques for system reliability — Procedure for failure mode and effects analysis. (Publication 812) IEC, Geneva, 1985.

IEC, International Electrotechnical Commission. Analysis techniques for system reliability — Procedure for fault tree analysis. (Publication 1025) IEC, Geneva, 1990.

ILO, International Labour Organisation. *Major Hazard control. A practical manual*. International Labour Office, Geneva, 1988.

INSAG, International Nuclear Safety Advisory Group. *Safety culture*. International Atomic Agency, Vienna, 1991.

ISO. Quality management and quality assurance standards — Guidelines for selection and use. (ISO 9000) International Standardisation Organization. 1987.

Johnson, W.G. *MORT Safety Assurance Systems*. National Safety Council, Chicago, 1980.

Karolinska Institutet, Department of social medicine. The World Health Organization. Manifesto for safe communities. 1st World Conference on Accident and Injury Prevention. Stockholm, 1989.

Kepner, C.H. & Tregoe, B. *The rational manager*. McGraw-Hill, 1965.

Kjellén, U. An evaluation of safety information systems at six medium-sized and large firms. *Journal of Occupational Accidents*, 5, 273–288, 1983.

Kjellén, U. The application of an accident process model to the development and testing of changes in the safety information system of two construction firms. *Journal of Occupational Accidents*, 5, 99–119, 1983.

Kjellén U. The deviation concept in occupational accident control– I, Definition and classification. *Accident Analysis and Prevention*, 16, 289–306, 1984.

Kjellén, U., Tinmannsvik, R.K., Ulleberg, T., Olsen, P.E. & Saxvik. B. *Sikkerhetsanalyse av industriell organisasjon — offshore versjon*. Yrkeslitteratur, Oslo, 1987.

Kjellén, U. & Larsson, T.J. Investigating accidents and reducing risks — a dynamic approach. *Journal of Occupational Accidents*, 3, 129–140, 1981.

Kjellén, U., Rydnert, B., Salo, R. & Östvik, R. Planering och genomförande av risk- och säkerhetsanalys i industrin. (SCRATCH-rapport), Nordforsk, Stockholm, 1983.

Kletz, T.A. HAZOP and HAZAN. Notes on the identification and assessment of hazards. The Institution of Chemical Engineers, Rugby, England, 1983.

Kliesch, G. ILO report by the chief of the occupational safety and health branch. *Proceedings XIth World Congress of the Prevention of Occupational Accidents and Diseases*. Stockholm, 1988.

Know, N.W. & Eicher, R.W. MORT user's manual. For use with the

Management Oversight and Risk Tree analytical diagram. EG&G Idaho Inc, Idaho, 1976.

Lees, F.P. *Loss Prevention in the Process Industries.* Vol 1 & 2, Butterworths, London, 1980.

Marnicio, M., Hakkinen, P. J., Lutkenhoff, S., Hertzberg, R. & Moskowitz, P. Risk Analysis Software and Databases: Review of Riskware '90 Conference and exhibition. *Risk Analysis,* 11, 545–560, 1991.

McAfee, R.B. & Winn, R. The use of incentives/feedback to enhance work place safety: A critique of the literature. *Journal of Safety Research,* 2, 7–18, 1989.

McElroy, F. (Ed.). *Accident Prevention Manual for Industrial operations.* (7th ed.) National Safety Council, 1974.

Menckel, E. & Carter, N. Tillbudsrapportering, en granskning av svenrk forskning. National Institute of Occupational Health, Sweden, 1984.

NCBS, National Central Bureau of Statistics. *Living conditions* Report no. 32, Working environment 1979. SCB Förlag, Stockholm, 1982.

NCBS, National Central Bureau of Statistics. *Statistical yearbook of Sweden 1992.* SCB Förlag, Stockholm, 1991a.

NCBS, National Central Bureau of Statistics and National Board of Occupational Safety and Health Statistics. *Occupational diseases and occupational accidents 1989.* SCB Förlag, Stockholm, 1991b.

Nertney, R.J., Clark, J. L. & Eicher, R. W. Occupance-use readiness manual — Safety considerations. Systems Safety Development Center, Idaho, 1975.

Nielsen, D.S. The cause consequence diagram method as a basis for quantitative accident analysis. (Report Risö-M-1374) Risö National Laboratory, Denmark, 1971.

Nielsen, D.S. Use of cause-consequence charts in practical systems analysis. (Report Risö-M-1743) Risö National Laboratory, Denmark, 1974.

O'Connor, P.D.T. *Practical reliability engineering.* (3rd ed.), Wiley, 1991.

Occupational Accident Research Unit. Program för Arbetsolycksfallsgruppen. Royal Institute of Technology, Stockholm, 1979.

Okrent, D. Alternative risk management policies for state and local governments. In *Risk evaluation and management,* ed. V. T. Covello, J. Menkes & J. Mumpower. Plenum Press, N.Y., 1986.

Petersen, D. *Human error reduction and safety management.* Garland STPM Press, New York, 1982.

Rasmussen, J. What can be learned from human error reports. In *Changes in Working Life,* ed. K. D. Duncan, M. Gruneberg & D. Wallis. Wiley, 1980.

Rasmussen, J., Brehmer, B. & Leplat, J. (Eds.) *Distributed decision making. Cognitive models for cooperative work.* Wiley, 1991

Rasmussen, J. & Jensen, A. Mental procedures in real-life tasks: A case study of electronic troubleshooting. *Ergonomics,* 17, 293–307, 1974.

Reason, J. *Human error.* Cambridge University Press, Cambridge, 1990.

Rouhiainen, V. The quality assessment of safety analysis. Technical Research Center of Finland, Espoo, 1990.

Rowe, G. Setting safety priorities — A technical and social process. *Journal of Occupational Accidents*, **12**, 31–40, 1990.

Saari, J. On strategies and methods in company safety work: From informational to motivational strategies. *Journal of Occupational Accidents*, **12**, 107–117, 1990.

Schaaf, T.W. van der, Lucas, D. A. & Hale, A. R. (eds.) *Near miss reporting as a safety tool.* Butterworth-Heinemann, Oxford, 1991.

Schelp, L. & Svanström, L. One-year incidence of occupational accidents in a rural Swedish municipality. *Scandinavian Journal of Social Medicine*, **14**, 197–204, 1986.

SCRATCH, Scandinavian Risk Analysis Technology Cooperation. *Sikkerhetsanalyse som beslutningsunderlag.* Yrkeslitteratur, Oslo, 1984.

Simon, H.A. *Models of man*, Wiley, New York, 1957.

SINTEF. Förslag till norsk standard, krav till risikoanalyser. SINTEF, Trondheim, Norway, 1991.

Suokas, J. On the reliability and validity of safety analysis. Technical Research Center of Finland, Espoo, 1985.

Suokas, J. & Kakko, R. On the problems and future of safety and risk analysis. *Journal of Hazardous Materials*, **21**, 105–124, 1989.

Suokas, J. & Rouhiainen, V. Work safety analysis. Method description and user's guide. Technical Research Center of Finland, Espoo, Finland, 1984.

Suokas, J. & Rouhiainen, V. Quality Management of Safety and Risk Analysis. Elsevier, 1992.

Sundström-Frisk, C. Behavioural control through piece-rate wages. *Journal of Occupational Accidents*, **6**, 49–59, 1984.

Swain, A. D. & Guttman, H. E. *Handbook of human reliability analysis with emphasis on nuclear power plant applications.* U.S. Nuclear Regulatory Commission, Washington DC, 1983.

Söderqvist, A. Rundmo, T. & Alltonen, M. Costs of occupational accidents in the Nordic furniture industry (Sweden, Norway, Finland). *Journal of Occupational Accidents*, **12**, 79–88, 1990.

Taylor, A. Comparison of actual and predicted reliabilities on a chemical plant. The institution of chemical engineers, IChemE Symposium No 66, 1981.

Taylor, J.R. Sequential effects in failure mode analysis. (Report Risö-M-1740) Risö National Laboratory, Denmark, 1974.

Taylor, J.R. A background to risk analysis. Electronics Department, Risö National Laboratory, Denmark, 1979.

Taylor, J.R. Evaluation of costs, completeness and benefits for risk analysis procedures. *Int. symp. on Risk and Safety Analysis*, Bonn, 1982.

Thiel,W. ISSA report by the chairman of the permanent committee on prevention of occupational risks. *Proceedings XIth World Congress on the Prevention of Occupational Accidents and Diseases*, Stockholm, 1988.

Vlek, C. & Stallen, J.P. Judging risks and benefits in the small and in the large. *Organizational Behaviour and Human Performance*, **28**, 235–271, 1981.

18 Index

The letter f after a page number denotes that page plus the immediately following page or pages.